Undocumented Migration

Immigration & Society series

Carl L. Bankston III, *Immigrant Networks and Social Capital*

Stephanie A. Bohon & Meghan Conley, *Immigration and Population*

Caroline B. Brettell, *Gender and Migration*

Thomas Faist, Margit Fauser, & Eveline Reisenauer, *Transnational Migration*

Eric Fong & Brent Berry, *Immigration and the City*

Roberto G. Gonzales, Nando Sigona, Martha C. Franco, & Anna Papoutsi, *Undocumented Migration*

Christian Joppke, *Citizenship and Immigration*

Grace Kao, Elizabeth Vaquera, & Kimberly Goyette, *Education and Immigration*

Nazli Kibria, Cara Bowman, & Megan O'Leary, *Race and Immigration*

Peter Kivisto, *Religion and Immigration*

Cecilia Menjívar, Leisy J. Abrego, & Leah C. Schmalzbauer, *Immigrant Families*

Ronald L. Mize & Grace Peña Delgado, *Latino Immigrants in the United States*

Philip Q. Yang, *Asian Immigration to the United States*

Min Zhou & Carl L. Bankston III, *The Rise of the New Second Generation*

Undocumented Migration

Borders, Immigration Enforcement, and Belonging

*Roberto G. Gonzales, Nando Sigona,
Martha C. Franco, and Anna Papoutsi*

polity

Copyright © Roberto G. Gonzales, Nando Sigona, Martha C. Franco, & Anna Papoutsi 2019

The right of Roberto G. Gonzales, Nando Sigona, Martha C. Franco, and Anna Papoutsi to be identified as Authors of this Work has been asserted in accordance with the UK Copyright, Designs and Patents Act 1988.

First published in 2019 by Polity Press

Polity Press
65 Bridge Street
Cambridge CB2 1UR, UK

Polity Press
101 Station Landing
Suite 300
Medford, MA 02155, USA

All rights reserved. Except for the quotation of short passages for the purpose of criticism and review, no part of this publication may be reproduced, stored in a retrieval system or transmitted, in any form or by any means, electronic, mechanical, photocopying, recording or otherwise, without the prior permission of the publisher.

ISBN-13: 978-1-5095-0694-1
ISBN-13: 978-1-5095-3180-6(pb)

A catalogue record for this book is available from the British Library.

Library of Congress Cataloging-in-Publication Data

Names: Gonzales, Roberto G., 1969- author. | Sigona, Nando, 1975- author. | Franco, Martha C., author. | Papoutsi, Anna, author.
Title: Undocumented migration : borders, immigration enforcement, and belonging / Roberto G. Gonzales, Nando Sigona, Martha C. Franco, Anna Papoutsi.
Description: Cambridge ; Medford, MA : Polity, 2019. | Includes bibliographical references and index.
Identifiers: LCCN 2019006575 (print) | LCCN 2019018019 (ebook) | ISBN 9781509506989 (Epub) | ISBN 9781509506941 (hardback) | ISBN 9781509531806 (pbk.)
Subjects: LCSH: Emigration and immigration--Government policy. | Immigration enforcement. | Illegal aliens--Social conditions. | Immigrants--Social conditions. | Social integration.
Classification: LCC JV6271 (ebook) | LCC JV6271 .G66 2019 (print) | DDC 325.73--dc23
LC record available at https://lccn.loc.gov/2019006575

Typeset in 11 on 13 pt Sabon by Servis Filmsetting Ltd, Stockport, Cheshire
Printed and bound in Great Britain by TJ International Limited

The publisher has used its best endeavours to ensure that the URLs for external websites referred to in this book are correct and active at the time of going to press. However, the publisher has no responsibility for the websites and can make no guarantee that a site will remain live or that the content is or will remain appropriate.

Every effort has been made to trace all copyright holders, but if any have been overlooked the publisher will be pleased to include any necessary credits in any subsequent reprint or edition.

For further information on Polity, visit our website: politybooks.com

Contents

Acknowledgments	vi
Introduction	1
1 Who Are Undocumented Migrants?	10
2 Theorizing the Lived Experience of Migrant Illegality	33
3 Geographies of Undocumented Migration	57
4 Immigration Enforcement, Detention, and Deportation	78
5 Undocumented Status and Social Mobility	98
6 Families and Children	121
7 Challenging Exclusion	144
Notes	164
References	168
Index	195

Acknowledgments

We would like to thank all of those who made this book possible. First and foremost, we want to express our appreciation for our editor at Polity, Jonathan Skerrett, whose patience and flexibility allowed us to finish the book during a time when undocumented migration has been front and center of the political debate.

Our efforts would not have been possible without the generous support – both financial and in kind – from our institutions and colleagues. In particular, we want to express our appreciation to Jim Ryan and Bridget Terry Long at Harvard University. We would also like to express our gratitude to Regina O'Brien, Deepa Vasudevan, and Leslie Molina for their helpful editorial assistance.

Ethnographic vignettes and citations included in the book come from research projects we have carried out individually over several years on undocumented migration. Some were published previously, in part or as a whole; others are published here for the first time.

Roberto wishes to thank Kristina Brant, Veronica Terriquez, Benjamin Roth, Felipe Vargas, Melanie Reyes, Victoria Villalba, Max Ahmed, Sayil Camacho, Edwin Elias, Ireri Rivas, Laura Emiko Soltis, Kathleen Sexsmith, Mary Jo Dudley, and Carlos Aguilar who were instrumental partners in the four waves of data collection for the *National UnDACAmented Research Project* (NURP).

Nando wishes to thank in particular: Alice Bloch and Roger Zetter with whom he worked on the *Young Undocumented*

Acknowledgments

Migrants project funded by the Paul Hamlyn Foundation; Vanessa Hughes for *Undocumented Migrant Children and Families in Britain*, funded by the Barrow Cadbury Trust; Elaine Chase, Jenny Allsopp and colleagues for the ESRC-funded *Becoming Adult: Conceptions of Futures and Wellbeing among Migrant Young People in the UK*; Heaven Crawley, Frank Düvell and colleagues for *Unravelling the Mediterranean Migration Crisis*, funded by ESRC and DFiD; and Laurence Lessard-Phillips for the ESRC-funded *EU Families and Eurochildren in Brexiting Britain*.

Thank you to the anonymous reviewers and to our colleagues and friends who encouraged and supported us at various stages in the preparation of the book. We are especially grateful to Sarah Horton and Ruth Gomberg-Muñoz for their careful read of an earlier draft. And thank you to our families – Sara, Joaquín, Julia, Robin, Matilda, Estefany, Ana Gloria, Tania, and Lena – for being the source of our inspiration and the support that underpins our work.

Finally, we wish to dedicate this book to those who traverse deserts, cross seas and rivers, and climb walls, facing incredible hurdles and sacrifices in search of protection and a better future for themselves and their loved ones. Their stories, words, smiles, and tears are an endless source of inspiration for us as researchers and human beings.

Introduction

A game of cat and mouse

At 225 meters long and five stories high, the former cruise ship, docked in the new port of Patras, in northern Greece, is a sight to behold. The ship dominates the entire port with its imposing body. Between ten and fifteen trucks are stationed outside. The drivers eagerly wait to board the *Cruise Europa*. Dozens of migrants can be seen observing the scene at a distance, eager to spot an opportunity to board the ship. Ticketed passengers and their vehicles will board much later.

Patras, a seaside medium-sized city, is situated in the northwestern part of the Peloponnese peninsula in Greece. The port was relocated out of town some years ago in order to decongest the city center from the massive trucks moving in and out of the port on a daily basis. For Vangelis, a thick-waisted employee of the Patras Port Authority, there was an urgent need to put an end to the chaotic scenes caused by migrants making a run for the trucks heading to Italy: "You cannot imagine the situation before. It was chaos: tourists, cars, trucks, and migrants all together. And the coast guard chasing them with cars and dogs. It was bad for the city and bad for tourism. And it was also dangerous for them [the migrants] too. We had many accidents and injuries."

The port of Patras is the main gateway of Greece to Europe. Due to the political upheaval and instability in the Balkans throughout

Introduction

the 1990s and the multiple land border crossings, Patras became the main route for both tourist and commercial transport in the region. Because the landscape around the new port is bare – the port is a vast surface of uninterrupted tarmac – there is nowhere to hide. This means that migrants making a run for the trucks are immediately in direct sight of the police.

The AVEX building, a disused wood-processing factory in the port area, is currently the makeshift home to some 380 migrants, mostly from Afghanistan, Pakistan, and the Maghreb. They sleep rough or in tents in the factory's derelict warehouse spaces that provide some protection from the winter's biting cold and the summer's scorching sun. They learn about this place mostly through their networks and through smugglers, as it is from there that they try every day to sneak into the trucks that are heading to Italy, and from there to other European destinations.

As one enters the port, one can spot migrants trying to climb the fence that separates the rest of the port from the departure gates, which is where all the trucks park waiting to board the ship. The steel fence is some two meters high and is reinforced with barbed wire on top, quite ramshackle in places from where people have climbed it.

This is an ordinary scene in the port, one that repeats three to four times every single day. Migrants show the scars and tears of these failed attempts on their bodies and clothes (see Jusionyte 2018 for discussion of migrant injury). They carry nothing with them except small bottles of water tied with thick rope around their waists or on their backs.

Attempts by migrants to board the trucks are not made in isolation. After someone gives a signal, groups of up to fifty young people make a run for the trucks. They then scale the fence, land on the other side and sprint towards the trucks. Numbers play a critical role in this endeavor: the more people participating, the better the chances that some will make it.

Truck drivers are also involved. Once the migrants come anywhere near the trucks, the drivers alert the coast guard by honking their horns. Patrolling coast guards in cars race in from the other side of the port, sending migrants scattering in all direc-

Introduction

tions. Migrants hope to avoid being run over by the speeding cars or bitten by the police dogs. Usually, when migrants are apprehended, the police keep them for several hours and then let them go again.

There is a strange familiarity between the migrants and the police. They seem to know each other. As night falls and all the trucks have boarded the *Cruise Europa*, the police approach the fence and announce the end of the game: "Finished for today. Tomorrow we do it again. Who's injured? Let's go to hospital."

The age of undocumented migration

International migration has risen in significance on the agendas of wealthy countries. According to the International Organization for Migration (IOM), the number of international migrants reached 258 million in 2017, an increase from 220 million in 2010 and 173 million in 2000 (IOM 2019). And, while undocumented migrants comprise only a small proportion of a country's immigrants, unauthorized migration dominates most discussions and perceptions of immigration (Jones-Correa and de Graauw 2013).

In response to the rise in global migration and its heightened visibility, many western nations have installed and aggressively fortified immigration controls on an unprecedented scale to deter migration and to punish immigrants. These measures constrain everyday choices as well as life trajectories for migrants. They also sow fear and anxiety in communities across the globe. At the same time, migrants residing in these countries form families, establish community connections, participate in local economies and governments, and pursue love, happiness, meaning, and political participation, just like their citizen neighbors.

Scholarship on undocumented migration often accords primacy to the nation-state in producing illegality, due to its power to define its relationship to citizens and noncitizens and to mobilize an enforcement apparatus to police the boundaries of membership. In many respects, we agree with this depiction. However, in this book we expand the analytical focus to include the interplay

Introduction

between different national and supranational configurations of "illegality." We argue that, while the production and experience of "illegality" are strongly shaped and determined by the state and state-based rules and regulations, they are more broadly framed by processes involving multiple states and international agencies and increasingly nested in multiple scales of governance. Furthermore, they are stratified by gender, class, and race.

This book explores state efforts to illegalize some forms of human mobility and the response of immigrants, their families, and their political and social allies to increasingly intrusive, repressive, and punitive measures of immigration control. In doing so, we draw a wide circle around undocumented migration to capture the international foundations and dimensions of "illegality." Through an examination of empirical examples in the United States, the European Union (EU), and the United Kingdom, as well as current scholarship, we trace how immigration policies and practices inform, influence, and impact decisions to migrate, the nature and length of migration journeys, and the experiences and opportunities of migrants in diverse host-country settings. At the same time, dynamics and political developments in receiving countries further complicate experiences of migration and settlement. To that end, we illustrate how migrants' everyday experiences are shaped by a range of laws and policies, from those explicitly targeting immigration and settlement of noncitizens to others that regulate access to public services, or regulate labor market more generally, but that nonetheless shape the possibilities and opportunities available to an individual with precarious immigration status.

From the journeys that migrants take to the lives they lead on arrival and beyond, this book aims to provide a comprehensive examination of the global, yet also local, phenomenon of undocumented migration. As such, we offer a triptych portrait of contemporary undocumented migration, which links: (1) the macro-societal processes that produce undocumented migration; (2) the shifting governance of immigration across different national borders and various locales; and (3) individuals' everyday experiences residing in host countries as undocumented migrants. This book primarily features the experiences of immi-

Introduction

grants in the United States, the United Kingdom, and Europe. As such, our examination does not, for example, capture the dynamics of migration between countries in the global South – like migrants leaving Myanmar for Thailand or Venezuelans fleeing political unrest and poverty for nearby Colombia – and of other wealthier countries like Canada or Japan. However, we draw on some of the major themes in the academic literature and in the popular discourse on immigration. Indeed, undocumented migration is now a global phenomenon. As the effects of armed conflict, environmental disasters and injustices, climate change, and economic inequality render large swaths of the planet uninhabitable, migrants seek refuge in countries of traditional settlement and those that have not been seen as long-standing immigrant destinations. This book is our effort to bring together a diverse set of issues and understandings about contemporary migration.

Our approach is driven by our interest in capturing the plurality of scales at which "illegality" is produced and experienced. In service to these goals, we have developed the concept of *illegality assemblage*. This is a term we use to describe the loose and dynamic system of laws and practices that transcend national borders and in which different interests and agendas find some kind of accommodation. We think of "illegality" as the product of a multi-scale and multi-actor assemblage that produces various configurations of rights, entitlements, constraints, and challenges in places in which migrants' lives unfold – thereby affecting every aspect of their lives as individuals and in families and communities.

About the book

We began the book with an ethnographic description of a modern-day game of "cat and mouse" to emphasize the diversity of structures, processes, and actors involved in migration projects across multiple state boundaries. Furthermore, this vignette illustrates our methodological and geographic foci. As ethnographers and qualitative researchers, more generally, our work examines the everyday experiences of migrants living in liminal, precarious,

Introduction

and clandestine statuses in the United States, the United Kingdom, and the European Union. Much of our work has also focused on young people of various non-legal statuses. Thus the empirical and theoretical material framing this book is reflective of our particular vantage points as scholars. Through this perspective, we aim to shed light on undocumented migration as a global and transnational process. We focus on how multiple nation-states contend with this phenomenon while also examining how the everyday experiences of individual migrants are shaped by illegality. By zooming in and out of global, national, local, and individual contexts, we animate the multiple dimensions that produce, reproduce, and contest illegality.

This book has been written during a period of incredible turmoil in our respective countries. On September 5, 2017, US Attorney General Jeff Sessions announced an end to the Deferred Action for Childhood Arrivals (DACA) program, an administrative action by the Obama administration to protect young undocumented migrants from deportation. The move followed similar actions to end protections for immigrants holding Temporary Protected Status (TPS) from six countries. The repeals of DACA and TPS were central to Donald J. Trump's presidential campaign promises to dramatically curtail immigration to the United States. Taken together, these actions directly impact more than one million immigrants living in the United States.

Trump's election and the ascent of populism in the United States came on the heels of an equally dramatic chain of events across the Atlantic Ocean. Following the Arab Spring uprising and civil unrest in North Africa and the Middle East between 2011 and 2018, more than two million people crossed the Mediterranean to Europe by boat without authorization. Over 17,000 deaths were recorded among people trying to cross the sea, and many more unrecorded deaths are thought to have occurred along land routes. In 2015 alone, at the height of Europe's so-called migrant/refugee crisis, more than one million arrivals were recorded in Italy and Greece, and 3,771 people died while crossing.

Lack of preparation in response to sea arrivals contributed to widespread moral panic among Europeans and the emergence of anti-immigration political movements in several EU member states.

Introduction

In the United Kingdom, fears of a migrant invasion played a major role in the referendum that saw 52 percent of British voters cast their vote to leave the European Union after more than forty years of membership. Overnight, three million non-British EU citizens were forced to grapple with a change that would dramatically reshape their futures. The European citizenship that had provided them the freedom to live and work in any of the EU member states since the early 1990s, albeit without the right to vote in national elections and referenda, was set to dissolve, casting their legal status and their lives in the United Kingdom into uncertainty. Similarly, about one million British citizens living in other EU member states saw their life and work prospects change dramatically.

A brief note on terminology: in this book, we mainly use the terms *undocumented* and *unauthorized* to refer to those migrants residing in destination countries without legal status. However, when referring to particular terms from the literature or from the media and, when trying to make a more specific point, we also employ terms such as *clandestine*, *precarious*, and *irregular*. We use the term *migrant* to denote those who are on the move or are settled outside of their country of birth, irrespective of their motivations for doing so, their goals, or their countries of origin. We nonetheless employ appropriate terms to denote people's different immigration statuses when necessary (e.g., asylum seeker, refugee, DACA beneficiary). Without conflating or ignoring the categorical and experiential differences between various groups of migrants (undocumented migrants, migrants with resident and work permits, refugees, asylum seekers), we wish in this way to highlight the permeability of the boundaries between different statuses. We acknowledge that language use as it pertains to the study of "illegality" is dynamic and often reflects the ever-changing nature of immigration laws, policies, and practices across different nation-states.

Outline of the book

This book highlights the dilemmas of undocumented migrants who find themselves on the wrong side of geopolitically and

Introduction

socially drawn lines, the vast majority of whom live in fear and uncertainty, attempting to forge lives for themselves and their families despite their undocumented status. As such, most undocumented migrants exist at the margins of society, fighting for their dignity with fortitude. Our aim is to delve into the concept of undocumented migration and the experiences of undocumented migrants by drawing upon and connecting theoretical insights and ethnographic and empirical data. We start by developing our theoretical framework on undocumented migration as a global and transnational phenomenon. In chapter 1, we introduce undocumented migration as an elusive social phenomenon that crosses national borders and increasingly attracts the attention of the international community. We consider who counts as an undocumented migrant, how one becomes undocumented, and why some forms of human mobility are categorized as "illegal." We discuss how this conceptually and empirically slippery category of mobility makes counting and measuring this group difficult. While the state's rules and regulations determine what counts as undocumented migration and who is to be considered an undocumented migrant, we also argue in this chapter that the production of "illegality" exceeds the borders of the state and also involves processes at the local, regional, and global levels that ultimately shape the concrete experiences of undocumented migrants. In chapter 2, we offer a theoretical framework for the study of undocumented migration and understanding the experiences of undocumented migrants. The chapter considers the impact of precarious legal status on migrants' identity and sense of belonging. Chapter 3 considers the international, national, and local dimensions and their interplay. State borders operate as a sorting and filtering mechanism for human mobility, in which "illegal travelers" and border guards perform a game of "cat and mouse." We then shift the focus to the subnational level; in particular, we consider the urban scale and how urban space attracts undocumented migrants and enables the emergence of new political subjectivities. Cities, we argue in this chapter, can offer a space for undocumented migrants to mobilize, build solidarity, and claim rights and belonging. In chapter 4, we consider the impact of immigration enforcement

Introduction

on undocumented migrants, focusing in particular on how they experience detention and the looming prospect of deportation and enforced destitution.

Chapter 5 shows how various contexts at departure as well as in the country of residence structure social mobility for undocumented migrants and how these migrants forge lives for themselves, despite their precarious existence.

In chapter 6, we explore the family experience of migration and the toll that separation of family members from one another takes on all their lives, especially on the well-being and future prospects of the youngest members of the family, including their emotional well-being and the opportunities for education and employment. In chapter 7, we conclude by considering approaches to contesting the concept and perception of "illegality" and link the experiential dimension of undocumented migrants to structural factors that have come to produce "illegality." We discuss examples of contestation and resistance by undocumented migrants, focusing in particular on the DREAMers in the United States, the *Sans Papiers* in France, and their counterparts elsewhere in Europe.

1

Who Are Undocumented Migrants?

Mohammad was 13 years old when he left Afghanistan in 2005. The Greek authorities found him twelve months later, on a beach, hungry and dehydrated after a 23-hour boat crossing in a dinghy with only one oar.

After both their parents died, Mohammad's eldest sister told him and his brother, Abdul, that there was no future for them in Afghanistan. Life in their home country had become too dangerous and those who had killed their father would come back for them. Soon after, Mohammad and Abdul joined two dozen young men and children on the back of a truck headed for the country's western border, near Iran. Once they reached the border, the group split up. Border guards were shooting at any sign of movement, and the group rationalized that it would be safer if they dispersed. Abdul told Mohammad to stay close.

The brothers walked for more than four hours in the dark until they reached the other side of the border. They arrived at a village where they stayed overnight in a house with a family of strangers. In the morning, they caught a local bus that took them to Tehran. In Tehran, they stayed a few days while the smugglers in charge of the next leg of the journey procured some forged passports for the group.

Eventually, another bus took Abdul and Mohammad to Urmia and, from there, to the border with Turkey. They were worried they would be easily spotted, but they managed to enter Turkey "dressed as business men." However, funds for the group were running low. Brought to Istanbul, the brothers and their fellow travelers – more than twenty altogether – were locked in a basement apartment, as Mohammad described, "not much bigger than a one-bedroom flat": "We were like mentally going crazy. People wanted to go out, but there was security outside. They had knives and guns. They were like, 'You have two choices: cross this door and you are dead or stay inside and live.'"

After three months of living in cramped quarters, the head smuggler informed them that in ten days the group would board a boat. "He took us to a coach and we traveled for five or six hours. Then we got to the sea and there was a big jungle with lots of animals like snakes and tigers. A massive jungle it was. 'That's the sea, the other side is Greece,' the agent said." But there was only space on the boat for one more person. Abdul insisted that Mohammad go without him, and that he would follow soon after. Mohammad did not want to leave his brother behind, but he had little choice. Abdul had the following words of advice for Mohammad: "Whatever country you go [to] please stay in education, don't make any troubles. Don't smoke, don't drink and . . . make me proud. Make your family proud." This would be the last time they spoke. The supposed second boat never came.

After the first two attempts at crossing to Greece failed, Mohammad found himself back in Istanbul standing purposeless in a park, without his brother and with no money left. There, a stranger approached him and asked if he was looking for work. After a drive of some hours, the car arrived at a farm somewhere in the middle of the countryside. For several days, Mohammad joined other migrants picking tomatoes. He endured countless hours of backbreaking work and was fed a meager portion of eggs at breakfast and a watery soup at dinner. After ten days, Mohammad demanded to be paid.

> So, I worked there for ten days and I said, "Now you have to give me some money." And the farmer said, "No, no, no, no. You work free for the rest of your life. Because there's no way you can get away from here. You're mine now; you work free." I thought this is really bad, I'm a slave. The next day I escaped from the home, but they took me

back, saying "This is the last time you escape [. . .]. Everyone knows you work for me. They will bring you back. I will call the police and they will send you back." And I said, "No, don't call the police, I don't want to go back to Afghanistan."[1]

Eventually, Mohammed managed to escape. He headed back to Istanbul and found an acquaintance who helped him to board a boat. His third attempt succeeded. From Greece, he moved to Italy, where he was impressed by the generosity he encountered among ordinary people. He even found a person who offered to pay for his train ticket to Rome. He then made his way to France and eventually managed to arrive in Calais, the port city on the English Channel. In Calais, Mohammad joined hundreds of other migrants and refugees in the informal encampment known as "the Jungle." There was not much to do there, other than wait for the right truck to hide in. It took multiple failed attempts and a few months to find a truck to take him to Britain. He carried out the final leg of the journey with an Afghan boy called Ahmed he had met in the Jungle. The boy was well educated and a gifted artist.

> He was like "Oh, Mohammad, do we need money?" I said obviously we need money. He said, "I know how to make money, trust me. I'm an artist I can draw." So, he went back to buy some brushes and stuff like that. He said, "Sit next to me here. Do you want me to draw you?" Many people saying yeah. And he would draw like exactly as they were, like very, very beautiful drawing. And people were giving him like 50 euros, 60 euros and they were like "Wow, that's a talent!"

Selling a few portraits by night, the two boys were able to afford food and clothes, a luxury not available to others in the Calais camp. Eventually the opportunity came to leave, and they reached England. Unfortunately, they were picked up by the police and, because they were minors, they were brought to social services. In this instance, being well educated and smart worked against Ahmed.

> I was the stupid one. Whatever they were saying to me I was answering like the way they wanted, and they thought maybe he is [a minor].

Who Are Undocumented Migrants?

But with [my friend], they thought he was older, that he was my older brother. [While in custody] he called me: "They say I'm eighteen," he told me. "It's more likely they will send me back to Afghanistan." So, I [told him to] run from here.

His friend absconded, avoiding any contact with social services. He kept in touch for a brief time but then disappeared.

Mohammad is now 23 years old and completing his master's degree in a city in the English Midlands. He has refugee status. The perils he endured during the journey – the forged passport, slave-like working conditions, countless nights spent sleeping without a roof over his head, the clandestine crossing of multiple borders, and the loss of his brother – were the price he had to pay to claim asylum in the United Kingdom: an undocumented journey was the only way to become a documented refugee in Europe.

For Mohammad, the clandestine journey and forged passport were the only means available to him as a young Afghan to find protection and security. Increasingly restrictive visa policies make access to international protection impossible to the inhabitants of some of the world's top refugee-sending countries. Reaching Britain, however, did not mean achieving asylum, at least not right away. The asylum process was long and convoluted, and there were several appeals before Mohammad was able achieve secure legal status and could begin to rebuild his life.

Unauthorized journeys

Mohammad's story is far from unique. Between 2014 and 2017, almost two million people entered the European Union on dinghies and unseaworthy boats. They endured a great deal of hardship during their migration journeys, sometimes marked by death. Over roughly the same period, nearly 14,000 people died crossing the Mediterranean. Many more unaccounted for died on land, from hypothermia along mountainous border paths, from lack of water and heat crossing the Sahara, from torture and violence in

Libyan illegal warehouses, and at the hands of predators specialized in targeting people on the move.

According to the UN High Commissioner for Refugees, the vast majority of those who boarded boats to Europe, 84.5 percent, came from four countries: Syria, Afghanistan, Iraq, and Eritrea. Most of these individuals are now lawful residents in Europe, with state recognition as refugees or with other forms of humanitarian protection. But the psychological and physical scars of such unauthorized journeys, of the people left behind or lost along the way, of the violence witnessed and experienced, of being treated like slaves – those will not go away, regardless of their future immigration status (Allsopp and Chase 2017).

Crossing international borders, by land or sea, can be a treacherous endeavor. Today's immigration controls rely on several modes of enforcement that operate at state borders, within the territory, and increasingly externally in other sovereign states.[2] Detention, deportation, and forced destitution have become normalized as tools of immigration management in traditional receiving countries, and they feed a booming migration enforcement industry (Andersson 2014; Gammeltoft-Hansen and Sørensen 2012; Vogt 2013).

In addition to the dangers inherent in unpredictable geographies, the journey exposes undocumented migrants to violence and danger. From official state responses such as incarceration and punishment to the abuses and violence perpetrated by border patrol agents, violence, injury, and death are oftentimes not merely side effects of border controls but are elemental to current militarized strategies (De León 2015; Jusionyte 2018; Slack et al. 2016).

Migrant women are particularly vulnerable (Schmidt and Buechler 2017). For Central American women crossing multiple borders en route to the United States, threats come in multiple forms – bandits, corrupt officials, travel companions, and smugglers (Brigden 2018). Often outnumbered by men and vulnerable to crossing guides (or smugglers), migrant women are at risk of kidnapping, assault, and rape (Simmons, Menjívar, and Tellez 2015).

But those who attempt to leave their native countries often face violence at home. Physical and economic violence often drives migratory decisions (Schmidt and Buechler 2017). As Simon McMahon and Nando Sigona (2018) argue, while most migrants and refugees are aware of the dangers associated with unauthorized crossings, many view the journey as the only thing left separating them from what they hope is a better and safer life. This point is evidenced by Clara, an Eritrean woman who migrated to Italy by boat. She was interviewed a few weeks after her arrival in Italy in 2015:[3] "We were very afraid on the boat. We could die. But at that point if you have lived the terrible things I saw in Libya and other countries, you do not care anymore about dying. It is almost better to die."

Undocumented migration in a changing world

The act of migration cannot merely be reduced to a set of choices made by individuals. Just as migrants embark on cross-border journeys, human mobility is heavily regulated through multiple interactions and different and often distinct governing contexts. In this current era, opportunities for authorized mobility have shrunk. And, for some, crossing without papers or obtaining a temporary visa only to overstay its terms is the only way to settle in destination countries. For many migrants, the act of migration is a precarious experience, marked by hardship, dehydration and hunger, exploitation, and even violence. Mohammad's story illustrates some of the complexity of undocumented migration. Along the journey, migrants confront a complex apparatus of smugglers, facilitators, border guards, and state actors. However, as Mohammad's story shows, undocumented journeys and irregular entries are sometimes the only way to apply for international protection and asylum.

But not all undocumented migrants enter receiving countries irregularly. Each year a sizeable number of undocumented migrants enter legally on a variety of visas with the intention of or by chance overstaying.

Undocumented migration does not exist in a vacuum. The

category "undocumented migrant" is only meaningful in relation to the contexts and circumstances that define it. What counts as undocumented migration and who is considered an undocumented migrant varies over time and space and is embedded in specific conditions, histories, and structures of power. As historian Mae Ngai powerfully argues, laws both reflect and constitute society. They naturalize and structure social relations (Ngai 2004: 12).

In many countries, concerns with undocumented migrants are prominently featured in political agendas. Persistent, misleading, and at times hostile media coverage of undocumented migrants can feed moral panic across diverse communities. Yet, despite renewed state efforts to increase patrols of national borders, to build longer and taller fences with neighboring countries, and to carry out swift mass deportation programs, undocumented migration persists around the world. Building a wall across one stretch of a country's border may have the short-term effects of satisfying immigration restrictionists and reducing migration in an area. But it may also increase crossings a few miles away while heightening risks and increasing deaths (De León 2015; Massey, Durand, and Malone 2002). And when migration to certain regions of the world begins to slow down, new flows emerge as migrants are redirected towards alternative routes.

Nations develop rules for formal membership that structure processes for citizenship and determine priorities for different types of immigrants. Based on these decisions, immigrants are assigned one of a range of different immigration statuses – e.g., citizen, lawful permanent resident, visa holder, refugee, asylum seeker, or unauthorized migrant – with varying levels of corresponding benefits to those in different legal categories and consequences for those on the wrong side of drawn lines. These acts of categorizing and labeling underpin the workings of the state and play a significant role in all governance matters, including the management of human mobility within and across borders (Gonzales and Sigona 2017). These designations also embed politics into the policy-making process, concealing political decisions behind the normalizing discourse of bureaucracy (Hastings 1998).

Rather than searching for inherent characteristics that might make an individual an undocumented migrant, we stress the function of immigration laws and regulations in defining legitimate and illegitimate forms of human mobility and attributing rights and entitlements. While illegality, then, is rooted in legal classifications, its power to limit social mobility and narrowly circumscribe everyday life lies not just in law but also in the discourses, politics, and practices which accompany the interpretation and implementation of laws (De Genova 2002). As Joseph Nevins (2002: 163) suggests, "[immigrant] categories become 'discursive facts' that inform how people interact [with] and perceive one another." These discursive facts, then, are imbued with meaning, either positive or negative, that generate perceptions of deservingness and membership. For many groups of noncitizens, this legal designation can connote a host of images that include lawlessness, criminality, and economic drain, and they thus become tools of domination and social control.

One current example of this process of labeling and its implications for categories of noncitizens is the current backlash against refugees and asylum seekers. In 2007, Roger Zetter, professor emeritus at the University of Oxford, identified two parallel processes that have worked in tandem to transform asylum governance and shift the public's sympathies away from those fleeing war, persecution, or natural disaster: (1) receiving states have generated a multitude of different types of immigration statuses and bureaucratic categories to manage forced displacement (for example, different types of temporary and non-renewable humanitarian protection statuses); and (2) there has been a move towards the *precarization* of rights and entitlements of those seeking protection, for example by making in the United Kingdom and other European countries a refugee status that was previously permanent now subject to a reassessment after five years and revocable in the event of changing political circumstances even after many years (Zetter 2007).[4] *Precarization*, in this sense, refers to the uncertain and changeable nature of laws, policies, and practices that bestow rights and entitlements upon those seeking refuge and protection. Although laws and policies seem concrete and stable, they are also dependent on multiple factors, including sociopolitical climates

Who Are Undocumented Migrants?

and attitudes. So, while refugees have long been viewed as worthy of societal acceptance, membership, and state-conferred benefits, an expanded set of categories deems some more worthy than others and provides rights for some but limits them for others.

In the United States, children and adolescents fleeing violence from Central America and arriving at the US–Mexico border without a legal guardian are designated "unaccompanied alien children." While this label provides these young people some protection from deportation, it does not bestow on them a formal legal status (Terrio 2015). More recently, flows of immigrants from Central America seeking asylum, among them families with small children, have been characterized by the government and media as an invading force, justifying the deployment of the US military to the southern border and prompting US President Donald J. Trump to approve the use of lethal force to prevent them from entering.[5] This punitive response is a clear escalation in how flows are managed across the border. But it is not a radical departure from recent actions.

This chapter describes the role of policy and governance in transforming large populations of noncitizens into undocumented migrants. In doing so, we trace the ways in which people become undocumented and offer some insights into challenges and limitations of efforts to measure and enumerate undocumented populations.[6] Recognizing how nation-states have shifted legal designations and definitions of migrants underscores how different forms of immigration statuses afford their bearers differential rights and entitlements. As such, how one becomes undocumented is a question of politics and policy.

Consequently, state categorizations of persons as citizens, lawful permanent residents, refugees, temporary foreign workers, subsidiary protection status holders, and undocumented migrants profoundly determine access to rights and other forms of resources. As such, immigration status designations accompany individuals as they attempt to enter the workforce, access labor protections, utilize social benefits, visit doctors, enter educational programs, participate in civil organizations, and encounter the justice system (Anderson and Ruhs 2010). For example, an undocumented status upon entry into the United States and the United Kingdom locks migrants into low-paid jobs without protections or benefits, excludes them from most medical and social services, and primes them as a target for government enforcement activities and anti-immigrant violence (Chavez 1991; De Genova 2005; Goldring and Landolt 2012; Gomberg-Muñoz 2011; Heyman 1995; Horton 2016).

Who are undocumented migrants?

Who are undocumented migrants? And how does one become an undocumented migrant? In a practical sense, an undocumented person can be described as someone who has no legal rights to reside in a particular country and limited entitlements to that country's polity. This definition, which perhaps seems straightforward, contains a number of elements that illuminate its complexity.

First, the term "undocumented" is qualified in relation to the particular country of migration and that country's legal framework. By outlining the criteria for entry and for residence, countries set the parameters for inclusion *and* exclusion. These criteria are influenced by a number of different factors. Notions of race and class, for example, often influence the *community of values* of a particular country and might then be enshrined in that country's immigration policies.

But while certain criteria may be written into the law at a particular time in history, these criteria can change over time and in response to changing political and social realities. Undocumented migrants can be laborers, refugees, students, and also highly skilled professionals with advanced degrees. They are loved ones, community members, neighbors, business owners, fellow parishioners, and students. While growing numbers of undocumented migrants cross international borders for economic opportunity and reside in host nations as laborers, others are displaced from their countries of origin by political turmoil, conflict, and extreme forms of poverty. However, undocumented migrants are also professionals who enter host countries on temporary visas only to stay beyond the specified time.

Who Are Undocumented Migrants?

Thus, upon further examination, the boundaries between the seemingly dichotomous categories of legality and illegality are less neatly defined and fixed than one might think. While public discourse often encourages the question, "*Who* is an undocumented migrant?," we argue that "How does someone *become* undocumented?" is the more critical question. At first glance, the answers to these questions may seem obvious, as the popular nationalistic refrain goes, "What part of 'illegal' don't you understand?" But the reality is far more complicated than what is presented in spirited political debates. *Becoming* undocumented, unauthorized, irregular, precarious, or clandestine in contemporary society is much more a process than it is a fixed state – a process that is seldom linear or unidirectional.

Individuals can become undocumented by crossing international boundaries or overstaying tourist or work visas. Many immigrants who today hold some form of legal status were once undocumented. Legal immigrants may be subject to losing their status due to crimes committed or relatively minor offenses. They can also become undocumented by virtue of changes in policy and through changing definitions of permanent residency, as may be the case for an estimated three million EU nationals who live in the United Kingdom in the event of a disorderly exit of the country from the European Union, so-called "no-deal Brexit."[7] People can also gain a more protected and legal status, only to lose it again. Such is the case of the more than 800,000 young people in the United States temporarily protected by the Deferred Action for Childhood Arrivals (DACA) program who were granted access to deportation relief, work authorization, and driver's licenses, but who stand to lose these rights if the program is revoked, and the more than 300,000 immigrants holding a Temporary Protected Status (TPS) who are also in jeopardy of losing their status due to the termination of the program.[8] And laws may define a range of legally permissible weekly hours for visa holders but deem anything in excess of those hours illegal (Kubal 2013).

Who Are Undocumented Migrants?

Estimating the number of undocumented migrants

The International Organization for Migration (IOM) estimates that there are 258 million international migrants worldwide.[9] This represents 3.4 percent of the world's total population.[10] Estimates of undocumented migrants are difficult to generate, but the same IOM report suggests that there may have been as many as 50 million "irregular migrants" as recently as 2009. While there is no universally accepted definition of undocumented migration, it is important to note that the phenomenon may refer to both the movement of people in an undocumented fashion and the number of migrants whose status may, at any point, be undocumented (Vespe, Natale, and Pappalardo 2017).

For researchers, authorities, and policy makers, assessing the number of undocumented migrants residing in host countries is a challenging task at both the national and global scale. This is due, unsurprisingly, to the clandestine nature of this population. Undocumented migrants rely on their ability to stay hidden from public view in order to remain in host countries. Therefore, efforts by researchers and government officials to enumerate populations of undocumented migrants tend to rely on estimates rather than actual censuses. These estimates often vary widely as they rely on different methodologies. For example, the Council of Europe estimated that 30 million people crossed international borders without authorization in 2002. By contrast, a 2004 estimate by the International Centre for Migration Policy Development (ICMPD) put the flow of irregular migrants at 2–4.5 million every year.

What's more, while some efforts focus on estimating the number of undocumented migrants residing at a given time in a particular place – what demographers refer to as *stocks* – others focus their efforts on tracking the number of undocumented migrants who cross borders – described as *flows*. Therefore, we share these estimates with a note of caution regarding both the ways in which these estimates are identified and how they are presented.

Among western democracies, the largest population of undocumented migrants resides in the United States. Jeffrey Passel, a

Who Are Undocumented Migrants?

leading demographer of migration, estimates the total number of undocumented migrants in the United States in 2016 to be 10.7 million, or 3.3 percent of the country's population that year.[11] Among this population, there are approximately, 7.8 million undocumented migrants in the US workforce. While they account for about 4.8 percent of the overall civilian workforce, they are overrepresented in farming (24 percent) and construction (15 percent). In addition, the vast majority of the country's undocumented migrants could be characterized as long-term stayers. In fact, about two-thirds of undocumented migrant adults living in the United States in 2016 had been living in the country for more than ten years, with a median stay of 14.8 years (meaning that half of them had been in the country at least that long). And, while they are dispersed throughout the United States, the majority of undocumented migrants are concentrated in a handful of states. Indeed, California, Texas, Florida, New York, New Jersey, and Illinois account for 58 percent of the country's undocumented migrants.

To be sure, undocumented migration to the United States is unrivaled among western democracies. However, the trend of massive undocumented migration, particularly from Mexico, is slowing down in recent years. The 2016 total actually represents a 13 percent decline from the peak of 12.2 million in 2007. And much of this decline has been driven by a decrease in undocumented migration from Mexico. Since 2012, net migration from Mexico to the United States has decreased. This means the number of undocumented Mexicans, as well as their share of the total, has been on a steady decline for the last several years. In 2016, the percentage of undocumented Mexicans dwindled to below 50 percent. While there were 5.4 million undocumented Mexicans living in the United States in 2016, there were as many as 6.9 million (or 57 percent) in 2007.[12] Interestingly, the total number from other nations, 5.2 million in 2016, remained unchanged.

As migration from Mexico has begun to decline, new trends appear to be emerging. In 2017, the US Department of Homeland Security (DHS) reported that more people overstayed their visas in 2017 than made unauthorized crossings that year. DHS reported that almost 702,000 people who entered the United States legally

Who Are Undocumented Migrants?

that year did not leave the country on time, compared to close to 362,000 who attempted to enter without authorization. These figures corroborate recent estimates suggesting that visa overstayers make up nearly 50 percent of all undocumented migrants in the United States. What's more, while typically thought of as a phenomenon involving Mexican and other immigrants from Latin America, growing shares of US undocumented migrants hail from countries in Asia. In fact, as numbers of undocumented Mexican migrants have been declining, the number of undocumented Asians has increased substantially. From 2000 to 2013, it increased by 202 percent.[13] Nevertheless, the majority of undocumented migrants in the United States still hails from Mexico. In fact, no other sending nation constitutes a double-digit share of the country's undocumented migrants.

In Europe, the most systematic estimate of undocumented migrants was carried out by the Clandestino Project in 2009, an interdisciplinary project funded by the European Commission, aimed to support policy makers in designing and implementing policies regarding undocumented migration. The project estimated that the number of undocumented migrants in the then 27 EU member states in 2009 was between 1.9 and 3.8 million.[14] For Germany, which hosts the largest share of migrants in the continent, the project estimated that between half a million and a million undocumented migrants resided in the country. A more recent estimate by the Clandestino Project for Germany puts this number in the range of 180,000–520,000 in 2014 – a much lower number, to be sure. But this estimate was carried out prior to recent mass arrivals of migrants by boats in southern Europe and the opening of the so-called Balkan route that brought large numbers of newcomers to Germany.

According to data from the United Nations High Commissioner for Refugees (UNHCR), more than 1.6 million people crossed the Mediterranean, with a spike between the springs of 2014 and 2016 attributable in large part to the conflict in Syria. Civil war and political instability in both Libya and Egypt were also drivers of mass migration. However, one should be careful not to mistake unauthorized entry in the EU with undocumented stays. In fact, the vast majority of those who crossed the sea without authorization subsequently applied for asylum in the EU and received some form of permanent or temporary protection. In 2017, according to Eurostat data (2018), 46 percent of asylum applications were approved, and an additional 34 percent were granted protection at appeal stage.

It is also worth bearing in mind that while many asylum applicants entered the EU by sea and without authorization, crossing the Mediterranean was not the only entry route into the EU for those seeking international protection, as the discrepancy between sea arrivals and asylum applications clearly shows.

In the United Kingdom, efforts to estimate the number of undocumented migrants have received increased attention since Jo Woodbridge's research (2005; see also Vollmer 2008), for the Home Office estimated that the undocumented population in 2001 was between 310,000 and 570,000, with a median estimate of 430,000.[15] A study by Gordon et al. (2009) reviewed and updated the Home Office figure, adding an estimate of UK-born children of undocumented migrants. This study produced a central estimate of 618,000 and a range of between 417,000 and 863,000 undocumented migrants at the end of 2007, of whom the largest single category is thought to be visa overstayers.

As these examples suggest, drawing accurate estimates of undocumented populations is difficult and can produce wide-ranging differences. In addition to problems created by ever-changing categories, another challenge in identifying a precise number of undocumented migrants is that many estimates (particularly those conducted outside of the United States) are based on the official statistics of immigration enforcement authorities (e.g., border apprehensions, asylum and regularization applications) rather than on an official census. Most large-scale surveys carried out by researchers do not ask about immigration status and locating undocumented migrants to survey is a difficult task, given their clandestine nature and their tendency to avoid authorities. And, in the case of Europe, there is the additional problem of aggregating numbers across the different EU member states as this involves the risk of double-counting those who have moved across states within the EU.

Measuring the global population of undocumented migrants is

not only technically challenging but also conceptually problematic. Demographers often lack reliable and comparable data on populations of hidden migrants in most countries. But even when such data are available, the complex nature of migrants' immigration status complicates efforts to measure it. Immigration statuses are often fluid, changing frequently due to reforms in the legal and policy frameworks that regulate entry, residence, and work authorization. For example, the enlargement of the EU with the inclusion of new members in 2004 and 2007 turned their nationals who were residing in the European Union without authorization into "EU citizens" and therefore entitled them to free movement within the EU almost overnight. In addition, there are also a host of in-between or temporary statuses that pose additional challenges for demographers. Take asylum seekers, for example. While they receive formal permission to legally reside in host countries during the application period, their status is ultimately contingent upon the outcome of their asylum application and possible appeals against negative decisions.

Measuring the number of undocumented migrants can also pose ethical problems for demographers, researchers, and others wishing to draw estimates. Those who lack legal immigration status are often forced to live a life undetected by the authorities or anyone who might expose them to the authorities. Additionally, political climates and socio-cultural attitudes towards immigrants in particular contexts can make it difficult for undocumented migrants to want to be identified. Importantly, for researchers collecting data on undocumented migrants, this means wrestling with the tension that exists between carrying out research and maintaining a responsibility to protect this vulnerable population, especially when laws, policies, and practices marginalize them.

A comprehensive approach: *the* illegality assemblage

International migration is a complex phenomenon, with a multitude of factors that create and sustain it. For this reason, we share the view that, when studying international migration, it is best to take a comprehensive approach that accounts for multiple factors explaining it (Bean and Stevens 2003; Massey et al. 1994). By this, we mean that people migrate due to social, economic, and political transformations. These transformations are often violent and highly disruptive, in part because they include the expansion of markets and geopolitical realignments. As such, decisions to migrate have to do with multiple and intersecting drivers: favorable or unfavorable policies; economic disparities among countries; global capital's reliance on labor and its recruitment; historic relationships between sending and receiving countries; and the experiences of success or failure of earlier migrants.

Although this set of migration theories helps us to understand the various and varied processes and structures that frame migration, it is still not yet able to serve as an umbrella for all states, groups, and circumstances. Refugees, for example, have been largely left out of the equation. For them, none of these theories adequately captures the complexity of their experience, from war and displacement to initial and oftentimes secondary settlement.

We argue that one must view undocumented migration as a complex phenomenon that is comprised of and motivated by an *assemblage* of factors: climate change, political instability, persecution, human rights abuses, war, transnational ties, draconian immigration policies, and lack of legal pathways to migration. While many immigrants cross international borders without government authorization, many others initially enter countries legally with visas and then overstay them or breach their conditions of entry. Others, by virtue of being born to parents who are undocumented in countries that deny birthright citizenship, are undocumented from birth. Still others fall into an undocumented status without ever realizing or see being undocumented as a necessary temporary trade-off in a broader migration trajectory. For some immigrants, going underground is the only way to avoid being sent to countries where they face persecution, prison, or even death. And yet, for many migrants, immigration controls, rather than reducing inflows, have had the effect of forcing immobility –

trapping undocumented migrants in receiving countries (Massey and Espinosa 1997; see also Carling 2002).

Migration has traditionally been defined and addressed from a state perspective. States are protective of their sovereign power to decide who can enter and reside in their territory. However, a growing awareness of the interconnectedness of migration flows has pushed states, particularly those that see themselves at the receiving end of international migration journeys, to seek cooperation with neighboring and sending countries to impede if not halt unauthorized migration. However, while undocumented migration is a global phenomenon – both because it involves the crossing of multiple international borders and because undocumented migrants can be found everywhere – not all states touched by it share the same agenda or interests.

Undocumented migration is shaped by the interplay of different national immigration regimes and the supranational pressures of neoliberal globalization to promote the free movement of goods and services across national borders while selectively restricting the movement of workers. Yet undocumented migration is primarily viewed as a national phenomenon insofar as it is the immigration policies of host countries that determine whether someone falls inside or outside the circle of membership. However, undocumented migration is *international* in that it often involves the crossing of multiple international borders, and it is *transnational* because decisions made at different points by migrants and various state and non-state actors can influence migratory projects and trajectories elsewhere.

To capture the plurality of scales at which "illegality" is produced, we offer the *illegality assemblage* as a concept that describes the loose and dynamic system of laws and practices that transcend national borders and in which different interests and agendas find some kind of accommodation. We view "illegality" as the product of a multi-scale and multi-actor assemblage that produces various configurations of rights, entitlements, constraints, and challenges in places in which migrants' lives unfold – thereby touching every aspect of their lives as individuals and in families and communities. The term "assemblage," first used in contemporary social science literature by French scholars Gilles Deleuze and Félix Guattari (1988), denoted "a multiplicity . . . made up of many heterogeneous terms [. . .] which establish[es] liaisons, relations between them, across ages, sexes and reigns . . . [Its] only unity is that of a co-functioning: it is a symbiosis, a 'sympathy.'"

The word, "assemblage" was chosen as a translation from the French term *agencement*, which suggests the act of putting things together without necessarily considering an end result. In English, an assemblage is both "a collection or gathering of things or people" and "the action of gathering or fitting things together,"[16] alluding to its contingent and dynamic character. Even though Deleuze and Guattari never fully developed and theorized it, the concept of assemblage has been applied widely in the social sciences, largely in history, anthropology, and more recently, in border and surveillance studies. The concept's usefulness derives from its utility in offering explanations and accounts that incorporate a multitude of heterogeneous items in different temporal and spatial arrangements.

Therefore, assemblage theory rids us of the need for a single underlying governing principle, thereby allowing us to come to grips with the complexity of undocumented migration. It affords examinations of complex social phenomena without reducing them to their bounded, structured, and stable components (e.g., actors, spaces, and networks). Saskia Sassen (2006) employs the concept of assemblage to capture global arrangements of territory, authority, and rights and to analyze how these have come together throughout history. The significance of Sassen's approach lies in the way in which the assemblage of territory, authority, and rights liberates her to construct narratives that avoid state-centered accounts and linear, temporal, and causal threads of reasoning.

Haggerty and Ericson (2000) previously argued for understanding and analyzing surveillance also as an assemblage that "operates by abstracting human bodies from their territorial settings and separating them into a series of discrete flows. These flows are then reassembled into distinct 'data doubles' [digital duplicates of our lives captured in data and spread across assemblages of information systems] which can be scrutinized and targeted for intervention" (2000: 606). Following this line of investigation, Mark Salter (2013) describes the global regime of mobility and circulation as an assemblage whose function is to manage human migration. Salter deconstructs notions of sovereign power, observing "a governmental mobile biopolitics that comes to manage circulation through the security techniques of inclusion, facilitation, and acceleration as well as exclusion, detention, and imprisonment" (2013: 12). This approach is particularly relevant for our analysis of "illegality" beyond the state in which we examine the interacting and counteracting norms, practices, values, rationalities, and actors that operate at different geographic scales which, by interacting with and counteracting each other (Delanda 2016), shape the contours of the life-world of undocumented immigrants today.

The role of national contexts in defining undocumented migrants

In most countries, a single narrative – or, in some cases, perception – dominates discourse on undocumented migration. It could be a story about a particular mode of entry into a country (e.g., unauthorized border crossings) or a stereotype about a particular nationality. In the United States, for example, the image of the undocumented Mexican migrant crossing the US–Mexico border without permission has become synonymous with the term "undocumented migrant," or pejoratively "illegal alien." However, contrary to these popular images, people who have overstayed visas have outnumbered those making unauthorized border crossings since 2007 (Gonella 2017). In fact, in 2014, two-thirds (66 percent) of those designated as undocumented were visa overstayers. What's more, migrants from every sending country, except Mexico and countries in Central America's Northern Triangle, are more likely to be overstayers than unauthorized border crossers (Passel and Cohn 2018). And the largest-growing segment of undocumented migrants in the United States hails from Asia. In fact, Mexicans represent a shrinking portion of the US undocumented population due to a decline of 1.5 million people between 2007 and 2016). Nevertheless, Mexico remains the largest sending country – at 5.4 million, Mexicans make up roughly half of the US total. Upon closer examination, it is not difficult to find diversity within larger populations of undocumented migrants in the United States, a diversity that belies convenient political depictions that paint them in broad brush strokes.

The case of migration to Italy offers an interesting portrait of immigrants' statuses changing over time. While Italy has long been viewed as a sending country, since the mid-1970s migration patterns have reversed, even as emigration continues to be a sizeable phenomenon. As of very recently, there were more than five million foreign nationals residing in Italy (IDOS 2018).[17] Interestingly, the majority of these individuals have at some point in time during their stay in the country been considered undocumented, often more than once and for protracted periods of time (Cvajner and Sciortino 2010). For many of these migrants, ongoing state-sponsored regularization campaigns allowed them to adjust their status and move from a precarious position, potentially subject to deportation, to a longer-term, more secure one.

Spain offers a similar example. The country also went from being a sending country to a receiving country in a short period of time. This change occurred soon after 1986, when Spain entered the European Union. As in Italy, it is common for immigrants in Spain to spend substantial and sporadic periods of time without formal legal status. Because their status is often attached to their employment, and given the instability of the labor markets in which they work and their related work cycles, immigrants often go in and out of status (Finotelli and Arango 2011).

Finally, in the United States, there are more than 400,000 immigrants from El Salvador, Haiti, Honduras, and seven other nations who have lived in the United States under a Temporary Protected Status (TPS) designation. The US federal government offered this status to some migrants fleeing dangerous and extreme conditions, such as natural disasters and civil war. Starting in the fall of 2017, the Trump administration began to end TPS for beneficiaries from Haiti, Nicaragua, El Salvador, the Sudan, and Honduras. On October 3, 2018, a federal judge in California granted a preliminary injunction that stopped the Trump administration from terminating TPS for Haiti, Nicaragua, El Salvador, and the Sudan (Jablon and Thanawala 2018). As a result, TPS holders from these groups retain their status while this ruling remains in effect. But their futures are uncertain.

These examples illustrate the importance of understanding undocumented status through the government regulations and processes that create these categories of legality and illegality. The various processes by which migrants become undocumented underscore the importance of policies and practices have at all levels – global, national, and local – in producing undocumented migrants.

Conclusion

This chapter has introduced some of the foundational theories, concepts, and ideas concerning undocumented migration as a social phenomenon and as a human condition with a set of experiences. In doing so, we have provided some key definitions of undocumented migration and undocumented migrants, including the various ways in which certain migration trajectories become undocumented. In this chapter, we have also highlighted the disjuncture between undocumented journeys and undocumented stays, while stressing that legal categories are often fluid due to individual and systemic factors. In this way, we have shown the complexities and challenges involved in the study of undocumented migration.

As exemplified by Mohammad's story at the beginning of the chapter, migrants are increasingly forced to take convoluted and fragmented journeys. They are also subjected to a series of obstacles and they have to overcome many hurdles and hardships during these journeys as well as navigating pathways to legality, inclusion, and integration. From dangerous border crossings to ever-changing and increasingly hostile immigration policies in receiving countries, migratory journeys and experiences are a complex area of study informed by many disciplines, including sociology, anthropology, geography, and economics. We hence illustrate the interplay of the causes and drivers of migration as these are set out in the main theories on undocumented migration overviewed briefly in this chapter. The multiplicity of interactions between factors prompts us to briefly introduce the concept of an *illegality assemblage* as an analytic lens through which to understand this phenomenon.

Finally, this chapter has outlined the main perceptions of who is an undocumented migrant and who may become one. It has provided rough estimates of the size of these populations worldwide, in particular in the United States and the European Union. It is no surprise that because of the evasive nature of undocumented migration, efforts at counting and measuring the phenomenon often encounter insurmountable barriers. For these reasons, we briefly problematized the challenges, methodologies, and ethics of counting vulnerable populations who often prefer to remain unseen. While enumerating undocumented migrants helps scholars and government agencies to understand and address the magnitude of migration, it does not necessarily capture the experiences or implications of becoming undocumented – the difficulty of settling in host countries and the challenges as well as opportunities involved in everyday life. Recent social science research in this area has produced dense knowledge of the relation between migration controls and the experiences of those negotiating borders, bureaucracy, and belonging. A lack of lawful immigration status can frame most aspects of migrants' lives, diminishing access to economic and social opportunities and exposing them to exploitation, victimization, and marginalization.

In the chapters that follow, we present an exploration of the experiences of migration and of living with undocumented status from multiple perspectives. In doing so, we draw on examples from current research as well as our own empirical research in the United States and the European Union.

2

Theorizing the Lived Experience of Migrant Illegality

Sergio was two years old when he, his mother and brother made a 1,300-mile journey from Mexico to Los Angeles, California. The first years in Los Angeles were especially difficult for the small family. Sergio's mother struggled financially with the piecemeal wages she earned working as a seamstress. In order to make ends meet, she sought out other work doing odd jobs here and there. While the extra work provided the family with additional income, it also significantly decreased the amount of time she was available for her children. Despite it all, the family still struggled to make a living off her wages.

As a result, Sergio's family bounced from home to home, often staying in other people's living rooms, dining rooms, and even in one family's garage. Moving so often without having a family home made it difficult for Sergio to establish long-lasting relationships, especially at school. After some time, Sergio's mother remarried. His stepfather was a lawful permanent resident and was a consistent provider for the family. Yet Sergio understood that the family's newfound sense of security and stability was still on shaky ground.

The inequalities Sergio and his family faced, were made concrete in different ways. Growing up, Sergio lived in nine different homes and seldom saw his stepfather, who was always working. Additionally, his mother was always tired when he got home from school.

In school, Sergio enjoyed playing handball and cracking jokes, which helped him make friends. Once Sergio turned sixteen, though, his life changed. A trip to the Department of Motor Vehicles to take the exam for his driver's permit turned Sergio's excitement into confusion when he was asked for his social security number. He did not know his social security number and had never before been asked for it. He went back home to check with his mother. She told him that he did not have one and that he did not have legal status. This revelation was "like a smack in the face." He said that it "really sucked, it really messed me up."

While, for many adolescents, a driver's license marks an important step towards adulthood, for Sergio the denial of this important rite of passage became a roadblock – one of many he would confront. School had never been a particularly affirming place for Sergio but the knowledge that he would be unable to join his peers in important milestones due to his immigration status had an immediate impact on his school life and beyond. He began to feel disillusioned and unable to see his future. He started acting out in school, skipping classes, and getting into fights. Ultimately, these incidents led to his expulsion during his sophomore year at high school.

Unlike many of the narratives and images used to describe high-achieving undocumented migrant students that have garnered much media and public attention, Sergio's story illustrates a more ordinary pathway – one that does not exude a narrative of success and high achievement. He was not involved in extracurricular activities, his bedroom walls were not lined with awards and certificates, and he was not at the top of his class, all qualities that make up the archetypal "DREAMer." Importantly, Sergio did not feel comfortable disclosing his status to his teachers and other school personnel, and thus he had few people to turn to or to support him during difficult times. He explained: "I just felt like nobody there cared about me. I was a troublemaker, and that's all they wanted to see. Nobody knew me."

Sergio struggled to find consistent work. But he was also afraid to do anything that might get him into trouble. His stepfather's status as a lawful permanent resident made him eligible to sponsor

Theorizing the Lived Experience of Migrant Illegality

Sergio and the rest of the family to adjust their immigration status. While this was an exciting development for the family, they did not have a clear timeline of when their immigration status would change. The then Immigration and Nationality Service (INS) faced a backlog and different community members confirmed the process was a long one.

Sergio could only wait. The uncertainty made him anxious. He refrained from working and driving, but that exacerbated his feelings of limbo.

The possibility of jeopardizing his chances of adjusting his immigration status by getting caught with "fake papers" left him with little option but to wait. Waiting did not help make things easier and in 2003 he lamented:

> I can't do anything that could help me. I'm stuck. That's the point, I'm stuck. I mean, with citizenship I could get a good job, I could finish my education in a good college. If you don't have papers you can't do anything. You can't move into a new area, into a good area. You can't buy a house. You can't buy a car. You can't do anything. That's what's holding me back; that's what's holding a lot of people back.

In the time between 2004 and 2006, Sergio's life took another turn. His family realized that the person they had hired to help with their immigration case was not an attorney but simply a notary public. This person had falsely represented himself as qualified to offer legal advice and services pertaining to immigration. In Spanish, the term *notario publico* is particularly problematic. While a notary public in the United States is only authorized to witness the signing of forms, the title in many Latin American countries refers to a person who has received the equivalent of a law degree and who is licensed to represent others before the government. The notary public took thousands of dollars from Sergio's family over the years, claiming to need it to process numerous forms. While the family gave the gentleman much of their hard-earned money, nothing was actually being done to move their case forward. Ultimately, they found out that this man had not only stolen from them but had also victimized many members of their community.

Sergio soon turned twenty-one and started a family of his own. He met a young woman and they had a child together, which meant that Sergio had to assume greater financial responsibility. He took a full-time job at a factory 30 minutes away from his apartment. While he knew that this was a risky choice, he could no longer wait for the INS to get to his immigration case.

At work, Sergio met a co-worker who lived near him and offered him rides to and from work. Unbeknownst to Sergio, his co-worker did not have a clean record. One evening as they were driving from work, local police pulled them over. The tattoos that covered his co-worker's bald head prompted police to profile the co-worker as a suspected criminal. As the officer began to question the young men, he saw drug paraphernalia in plain sight. The officer asked them to get out of the car and during the search he also found a homemade explosive device. While Sergio was unaware of the contents of the car, he was still charged as an accomplice to a federal crime and ordered to serve a three-year prison term.

Sergio finished his prison sentence in 2009 and was deported to Tijuana, Mexico. He was 25 years old in a country he knew very little about and had not returned to since his departure as a toddler. He was forced to start from scratch. For the next two and a half years, Sergio was homeless off and on while he worked odd jobs to make ends meet. He became depressed but also realized that he was more resilient than he had imagined. During his time in Tijuana, he met an older woman who rented him a room in her apartment above a loud bar. Sergio recalls those two and a half years as the toughest years of his life.

In 2012, Sergio returned to Los Angeles. He disclosed few details about his return but he was now more cautious of the authorities. He tried to stay "under the radar." Eventually, he found work in an assembly plant a few miles from his home and he chose to walk to work. Ensuring that he avoided arrest and deportation was crucial now that he had two children. As a deportee, if Sergio finds himself apprehended by Immigration and Customs Enforcement (ICE) he risks a mandatory 10–15-year prison sentence.

Sergio's lack of immigration status has had a cumulative effect on multiple aspects of his life. Ultimately, his deportation forced him further into the shadows, making his world even smaller. Sergio's story illustrates some of the dimensions of the undocumented life. While much of his childhood was shaped by his mother's undocumented status, his own undocumented status came into sharp focus during his adolescence.

For those like him, navigating daily obstacles brings constant challenges. Immigration enforcement extended to community spaces narrowly circumscribes everyday life and marks their outsiderness. Yet, over time, they start families and grow roots in their communities. Particularly for children, these experiences can provide a sense of membership. Ultimately, though, migrants must contend with lives under stress, defined both by their belonging and their exclusion.

Recent debates around citizenship have wrestled with questions of membership and what it means to reside within a community without legal authorization. Is community presence sufficient grounds for asserting membership? Or does one first need to be recognized as a member? Recent trends in globalization, human rights, and multiculturalism have questioned the enduring power of citizenship, while many western democratic states have ramped up enforcement activities in an effort to exclude, surveil, and expel unwanted immigrants.

Legal scholar Linda Bosniak (2006) has drawn distinctions between what she refers to as citizenship's hard boundaries and its soft interior. Accordingly, expressions of citizenship's hard boundaries refer to the formal rules and practices employed to maintain boundaries. Examples include physical or territorial borders and the meanings attached to these borders through formal laws and policies that give value to citizenship, whether one falls inside or outside national boundaries.

However, within nation-states, there are soft interiors where social relationships are formed, links are developed, and where migrants practice everyday acts of membership through their labor and their community participation. Citizenship's soft side, then, alludes to the breaches that lie between the outer workings of immigration enforcement and the legal frameworks of citizenship, leaving room for varying forms of inclusion (Bosniak 2006). These contrasting depictions enlighten us about the dual and complex nature of citizenship. On the one hand, legal citizenship defines those who have a legal mandate to remain in and move about the nation-state. Citizens are afforded rights, privileges, and opportunities. On the other hand, social and cultural citizenship delineates the ways in which people – documented and undocumented – forge their own paths and make claims to be included. Social and cultural citizenship zeroes in on the more personal and the daily practices that establish community, networks, and ties. Simply put, today's migrants, citizens or not, can be removed from spaces and denied privileges and rights. But they can also experience belonging and make claims to informal forms of membership in the polity. These two conceptions of citizenship also map onto two important theoretical perspectives: migrant illegality and experiences of belonging.

Illegality in context

National borders are far from impermeable. Some might argue that this impermeability is by design. The filtering function of borders works in tandem with immigration policies to produce a migrant workforce that is precarious and exploitable in both sending and receiving countries, restricting the mobility of the majority to places where global corporations and local employers can pay minimum labor costs. For the few who succeed in reaching wealthier countries, their life is precarious as normal wage structures do not apply to them. Immigration controls often make the act of migration difficult, costly, and dangerous (see Massey, Durand, and Malone 2002). As a result, those who do make it to receiving countries are often too wary to risk repeat trips and are keen to stay hidden. As such, their precariousness subjects them to narrowly circumscribed everyday lives and labor exploitation (De Genova 2005; Gonzales 2016; Gonzales and Sigona 2017; Willen 2007).

While "illegality" may be experienced in different contexts and in relation to a range of behaviors, it is in the context of migration that it becomes a truly existential condition (Menjívar and Kanstroom 2013). Rarely outside this context do people consider themselves "illegal." Anthropologist Nicholas de Genova (2002) has analyzed illegality in depth as a social condition determined by a particular legal relation to the state, rather than as an intrinsically natural and homogeneous category of people. He argues that by understanding lawmaking as the creation and maintenance of nation-state and nationalism, one can recognize that its main aim is to mediate social tensions, antagonisms, contradictions, and crises, particularly regarding migrant labor. Viewed through this prism, it is evident that the distinction between legal and illegal subjects has been mainly deployed to regulate migrant labor through stigmatization and enforced vulnerability.

Migrants participate in many different spheres of society. They engage in family life, form social ties, attend churches and other institutions, and hold jobs. But the state, via immigration law, often occupies a dominant position in migrants' lives (Menjívar 2006: 999–1000). In this chapter, we shift our focus to the various ways in which migrant "illegality" has been theorized – in terms of how it is both produced and experienced. In effect, the emerging portrait of contemporary undocumented migration is multifaceted: it is one which links macro-societal processes with migrants' everyday experiences of "illegality."

Since the late nineteenth century, a series of policies and practices has constructed and subsequently modified categories of exclusion from nation-states, deepening the division between undocumented migrants and their legal counterparts (Bosniak 2006; Coutin 2000; De Genova 2002; Hagan 1994; Ngai 2004). While questions of immigration status have scarcely been present in studies of immigrant assimilation, there is a burgeoning social science literature that has sought to better understand unauthorized migration and the experiences of undocumented migrants (De Genova 2002).

Over the last several decades, these studies have been carried out by an interdisciplinary group of sociologists, demographers, and anthropologists examining border enforcement, the social and economic costs of migration, labor-market effects, gender and other forms of stratification, and health and welfare (for a review of these studies, see Donato and Armenta 2011). Early research in this area focused on the experiences of undocumented migrants in particular labor sectors (Burawoy 1976; Bustamante 1976; Portes 1978). Then research shifted to encompass relationships between sending communities and migrants' experiences of settlement (Massey et al. 1990). Soon after, having developed sophisticated means for drawing estimates of the size of the population, demographers carried out quantitative analyses to assess labor-market effects on undocumented workers (Bean, Telles, and Lowell 1987; Warren and Passel 1987). Furthermore, throughout the late 1980s and the 1990s, ethnographic research began to explore undocumented migrants' everyday experiences (Chavez 1991; Delgado 1993; Hagan 1994; Rodriguez 1987).

More recent work in the area of "migrant illegality" has explicitly examined the role that state structures and practices play in framing the lives of groups of individuals who fall outside the bounds of formal membership (those who are legally constructed as "illegal") and in enforcing and sustaining these groups' outsider status. Scholars working in this area argue that illegality is more than a juridical status; it is also a sociopolitical one (Coutin 2000; De Genova 2002). This perspective focuses on illegality as historically and legally produced and situates undocumented migrants within a broader framework of a global capitalist economy (Goldring, Berinstein, and Bernhard 2009; Menjívar and Kanstroom 2013). From this vantage point, illegality is a dynamic, sociohistorical process rather than a static concept. It is both produced *and* experienced (Chavez 2007; Gomberg-Muñoz 2011; Horton 2016; Willen 2007).

Illegality as a produced phenomenon

In her work on immigration policy in Spain and Italy, sociologist Kitty Calavita theorized the production of illegality as an ongoing process of "irregularization" (Calavita 1998). In her study, Calavita documented a process through which state efforts to integrate immigrants in the 1980s directly contradicted provincial policies and administrative requirements that granted workers a form of temporary legal status while they were working. During

Theorizing the Lived Experience of Migrant Illegality

off-work cycles, these migrants had no legal rights; they were ineligible for housing and were denied many government services. By welcoming immigrants as workers but denying them permanent residence, the governments of Italy and Spain restricted these groups' ability to settle legally and comfortably. As a result, migrants' marginality was socially constructed and regularly reconstructed, despite the countries' emphases on social integration. Here, immigration laws, rather than regulating immigration, worked to control and marginalize immigrants themselves.

Similarly, Mexican migration to the United States has taken place within the historical contexts of US labor shortages. America's immigration laws today are largely the result of a history of intricate and calculated interventions – strategies, tactics, and compromises – aimed at filling these intermittent shortages through continuous flows of flexible, pliant labor across the US–Mexico border (De Genova 2002; see also Bonefield 1994, 1995; Holloway 1994, 1995). These deliberate interventions have brought about an active process of inclusion of Mexican migrants in the labor force, but with only limited opportunities for these workers to be absorbed into the legal labor market and into the cultural fabric of the communities in which they settle (Calavita 2005; Ngai 2004). These processes have given rise to increasing numbers of low-wage laborers who meet the needs of the economy but do so without the protections and privileges of formal legal status. Working in tandem, immigration laws and labor-market demand have produced and sustained a legally marginalized surplus of workers exposed to conditions of extreme vulnerability and excessive and extraordinary forms of surveillance and policing (De Genova 2002: 424).

Nicholas De Genova's seminal work on migrant "illegality" has convincingly moved the analytic focus away from migrants as bearers of illegality to a more deliberate study of the mechanisms that produce and sustain "illegal subjects" (De Genova 2002). According to De Genova, the legal construction of migrant illegality is achieved through the development of a profound sense of deportability among members of immigrant communities (De Genova 2002). The experience of undocumented migrants is defined by the *possibility* of being removed. In other words, while not everyone is deported, the threat of deportation sustains migrants' vulnerability and docility. Increasingly, state exclusionary measures and enforcement efforts narrowly circumscribe the daily lives of undocumented migrants. By denying these individuals access to a broad range of services and pursuits, and at the same time policing public spaces, the state criminalizes common daily tasks such as driving and working, further constraining what is legally possible. The external forces that produce "illegality" also profoundly shape migrants' individual and collective experiences of being in the world, resulting in a state of forced invisibility, exclusion, subjugation, and repression (Coutin 2000: 30).

In recent years, a multitude of practices have been displaced from immigration authorities to individuals who ordinarily have little or no connection to immigration enforcement (e.g., office clerks in public primary, secondary, and post-secondary schools, housing staff, and welfare benefits administrators). These "frontline" workers (Lipsky 2010) are, nonetheless, empowered to make decisions that shape immigrants' access to services and resources. Often, decisions are tied to subjective notions of deservingness for which immigration status operates as a critical variable. This policing of public and private spaces has exacerbated immigrant communities' sense of ever-present vulnerability (Coutin 1993). By extending laws targeting immigrants to include the jurisdictions of local and state officials, the government has effectively expanded its "enforceable territory" while shrinking migrants' "autonomous zones of safety" (Coutin 1993). Undocumented migrants have been criminalized as "illegal" and, as such, subjected to ongoing and excessive forms of policing (Coutin 1999). Alongside these enforcement measures, they are also regularly denied fundamental rights and access to social services. What's more, they are relegated to an uncertain and marginal status, often with few protections or avenues of mobility.

The condition of illegality

Scholars studying migrant illegality have also identified central themes in the lived experiences of undocumented migrants. This

Theorizing the Lived Experience of Migrant Illegality

work has shone a bright light on the embodied, legal implications, and social processes that animate the experience of illegality. Importantly, this scholarship, which is primarily qualitative and ethnographic in nature, emphasizes how constructs like time and space are reconfigured by the condition of illegality.

Socio-legal anthropologist Susan B. Coutin argues that the combination of legal regulations and the threat of deportation render undocumented migrants "legally non-existent" – that is, legally outside the United States, while physically present within the country (Coutin 2000). She outlines several dimensions of "non-existence": migrants' reality is confined to only that which can be documented; their presence is temporalized, such that the possibility of regularization depends on the accumulation of continuous, verifiable (documentable) "illegal" residence; certain kinship ties are nullified by immigration policies; concealment is required ("enforced clandestinity"); and mundane activities such as working, driving, and traveling are transformed into illicit acts (Coutin 2000: 30–3). To this list, De Genova adds an "enforced orientation to the present," conditioned by uncertainties arising from the possibility of deportation, which inhibit undocumented migrants from making many long-term plans (Carter 1997: 196; De Genova 2002: 427). As one can see, these dimensions of undocumented life are tied up in contradictions, they fracture relationships and they orient undocumented migrants to a cruel and unpredictable bureaucratic system.

Sociologists Cecilia Menjívar and Leisy Abrego introduce the term "legal violence" to connect structural conditions with migrants' everyday lives (Menjívar and Abrego 2012). Increased immigration raids, apprehensions in community spaces, and mass detention and deportation associated with recent immigration enforcement efforts engender a form of violence inflicted upon individuals and families. These laws combine with various institutions to restrict immigrants' everyday lives.

In highlighting the subjective dimensions of migrant illegality, Sarah Willen (2007) has invited scholars to view illegality beyond a juridical status and a sociopolitical condition in order to understand how illegality also creates "particular modes of being-in-the-world." In other words, illegality is an experience that migrants have in common. In doing so, Willen conceptualizes these experiences as a form of "politically and socially abject status" and draws attention to the ways in which migrant illegality operates as "the catalyst for particular forms of abjectivity" (2007: 11). This approach, she argues, "challenges us to consider not only the underlying historical, ideological, and political economic factors that led to the radical reconfiguration of the condition of migrant illegality, but also its deeply significant experiential, embodied, and sensory dimensions as well" (2007: 13).

Migration scholars Roberto G. Gonzales and Leo R. Chavez (2012), in turn, elaborate the concept of abjectivity, suggesting that while undocumented migrants are expelled from the national body, they are also included as "others" who remain subject to the politics of citizenship. Drawing on the work of Agamben (1998) and Foucault (1979), they examine the ways in which disciplinary powers shape the social and material lives of undocumented migrants by: (1) drawing direct connections between their life opportunities and surveillance strategies, identification requirements, law enforcement practices, and national (and international) policies; and (2) by showing how undocumented migrants learn to internalize self-understanding and self-disciplinary practices that make concrete their undocumented status and oppressive conditions. Undocumented status, then, combined with policies and practices that restrict everyday life and discourse that positions them as undeserving members of society, pushes them further to the margins of society where they must carry out their everyday lives and avoid a growing range of actors who have the power to subjugate their lives even further.

This *experience* of illegality is felt at the level of everyday life and in the bodies of migrants. Immigration controls within communities produce a heightened awareness among migrants that their everyday life is surveilled, patrolled, and policed. This knowledge shapes daily perceptions of uncertainty, danger, and the threat of deportation. According to Willen, this heightened sense of awareness produces specific kinds of fear and anxiety that often have physical as well as emotional effects – what Willen aptly terms the "embodiment" of illegality. Ultimately, the complicated realities of everyday life in the shadows have tremendous implications for physical, mental, and emotional well-being.

However, in spite of policies that define undocumented migrants' presence as illegal and constrain their daily lives, these men, women, and children integrate themselves into local communities, cities, and eventually polities. They engage in community activities, form relationships with friends and neighbors, and start families. These everyday experiences can buffer the negative implications of exclusion and bolster a sense of belonging.

Understanding migrant illegality through a sense of belonging

In contrast to the experiences of migrant illegality, a different, equally compelling lens draws attention to the ways that undocumented migrants lead their everyday lives, form relationships, and participate in local communities. This perspective focuses on the people located within nations and their everyday practices. As such, it stresses insiderness over outsiderness and examines how undocumented migrants' participation in the broader community provides them with certain claims to membership. Research in this area has developed into an interdisciplinary subfield within citizenship studies. This work on informal modes of belonging provides a corrective to the limitations of dominant discourses of membership that overemphasize formal, legal immigration status as a prerequisite to an individual's ability to assert a claim to belong.

A growing body of literature in psychology and sociology builds on the concept of belonging (Yuval-Davis 2004). Social psychology has sought to understand individuals' need to conform to groups and how their interpersonal relationships are affected by their membership (or lack of membership) as well as by their position within groups. Sociological theory, in contrast, has long focused on the various ways people belong to groups, communities, and societies, and the effects on individuals of social, economic, and political dislocations (e.g., migration, globalization, industrialization, and gentrification). Australian social geographer Elspeth Probyn describes the ways in which belonging is about both "being" and "longing" (Probyn 1996). It is a concept commonly tied to membership: to be a member of a group or community; to be a resident of or closely connected with a particular community; or to be rightly placed or classified within a community. Indeed, people "can 'belong' in many different ways and to many different objects of attachments. These can vary from a particular person to the whole of humanity, in a concrete or abstract way; belonging can be an act of self-identification or identification by others" (Yuval-Davis 2006: 199). Belonging manifests in individuals' everyday practices and materially through their spatial environments. As such, belonging is also associated with past and present experiences and with memories connected to a particular place (Fenster 2005: 243).

While social scientists have long studied the *structure* of communities, social psychologists in the 1970s began turning their attention inwards, towards questions of the *experience* of the community. Community psychologist Seymour Sarason believed that healthy communities exhibit an extra-individual quality of emotional interconnectedness that individuals enact in their collective lives (Sarason 1974: 6). For Sarason, this sense of community was rooted in the notion that individuals feel part of a larger, dependable, and stable structure. Hence, perceptions of community are shaped by a sense of similarity and interdependence with others (Sarason 1974: 157). While we typically think of communities in terms of the geographical aspects of neighborhoods, towns, and cities, this *sense* of community has to do with the nature and quality of relationships in a specific geographical space (Gusfield 1975). Thus one's sense of community could imply a sense of belonging to a particular area, but it could also include the character of human relations, without any reference to a particular location (Gusfield 1975: xvi). While these two usages are not mutually exclusive, proximity or shared territory cannot by itself constitute a community; the relational dimension is essential. A *sense* of community, then, includes membership, influence, fulfillment of needs, and a shared emotional connection. This suggests that individuals need to feel a sense of belonging to communities, not solely for material reasons but also for their own emotional and mental well-being (McMillan and Chavis 1986: 6–7). Accordingly, community membership is understood as a feeling that someone has invested part of themselves to become a member and therefore has a right to belong. And while membership in communities has boundaries (and those boundaries provide positive benefits to communities), it is often acquired through intimate cultural knowledge rather than through formal rules.

Much of the immigration literature on belonging has asserted the aforementioned premises of social and emotional attachments as forms of membership. Immigrants often feel part of a community because of sentiment influenced by social relationships and cultural beliefs and practices (Coutin 2003). Over the last several years, debate about the definition of citizenship has intensified, with some questioning whether it should even be understood in relation to the nation-state.[1] Scholars have coined phrases for alternative forms of citizenship: "global citizenship," "transnational citizenship," "post-national citizenship," "social citizenship," and "multicultural citizenship" (see Bosniak 2000, 2006). Some of these conceptions propose models of societies in which different populations participate through group membership, rather than on the basis of individual rights. There are seldom complete overlaps between the boundaries of a nation and the boundaries of the population that lives in that nation, which gives credence to these various and broad definitions of citizenship (Yuval-Davis 2006).

Drawing distinctions between legal forms of citizenship that determine "the full exercise of legal rights," on one hand, and participatory forms of citizenship that denote an "effective presence in the public sphere," on the other, scholars have revealed these routine manifestations of citizenship as both participatory and local in character (Balibar 1988: 724; cf. Reed-Danahay and Brettell 2008). T. H. Marshall defined citizenship not just as community membership but also as including rights and responsibilities. These more democratic articulations of citizenship have more to do with the community participation of citizens as social and political actors than with laws endowing them with the rights to do so. It is the *being* of citizenship that is emphasized over the *becoming* (Castles and Davidson 2000). Asserting cultural forms of citizenship involves agency. Sociologist Nira Yuval-Davis draws an important distinction between belonging as an "emotional sense of home" and "the politics of belonging" (Yuval-Davis 2006). For her, these politics are not just about the continual work the state puts into maintaining and reproducing community boundaries. The politics of belonging are also about the many challenges to those boundaries by members within the community.

The concept of "cultural citizenship" allows immigrants to be recognized as legitimate political subjects, and in that sense, to be also considered citizens. Local articulations of citizenship become predicated not on legal recognition and the granting of rights but on commonplace contributions to the community that confer the right to basic protections and respect. These notions highlight different ways of being a citizen that are connected to the experience of belonging. In their conception of cultural citizenship, anthropologists Renato Rosaldo and William V. Flores highlight local, informal articulations of membership. This takes into account "vernacular notions of citizenship," the broad range of activities and relationships in everyday life through which disadvantaged groups claim space, and eventually rights, in society (Rosaldo 1994, 1997; Rosaldo and Flores 1997). Contrary to claims that globalization produces homogenization (at the social, cultural, and political levels), cultural citizenship suggests difference can also be created through global transformation. It asserts that the civil rights, basic protections, respect, and recognition granted to citizens (i.e., the benefits of citizenship) can be extended to all residents who contribute to society. As Rosaldo and Flores point out, immigrants' labor contributes to the economic and cultural wealth of nations and their presence and community participation add to a country's social and cultural landscape. Through their expression and creation of alternative cultural and political spaces (e.g., public cultural events, organizations representing the interests of immigrant workers), they affirm their culture, identity, and rights as workers and local residents (Stephen 2003).

Undocumented migrants' claims to community, cultural citizenship and belonging push back against dominant conceptions that cast them as outsiders.[2] The second-class status of Mexican migrants in the United States is reinforced by a racial history that has repeatedly defined persons of Mexican descent as racially inferior and biologically best suited for agricultural labor (Molina 2014). Unlike the concept of legal citizenship, which designates and labels undocumented migrants as "illegal," cultural citizenship moves citizenship from the federal legal sphere to a more local and interpersonal articulation. By rooting citizenship in the everyday practices of immigrants, this alternative conception underscores and reaffirms their contributions outside of legal frameworks.

The thin boundary between illegality and belonging

Migrants seldom experience undocumented status as an isolated social marker, through the framework of community belonging, or as an embodied constraint (Kubal 2013: 11). Rather, their experience of "illegality" is shaped by various spaces and through many different types of interactions: tight labor markets that are prone to abuse and harsh conditions but where they meet other co-ethnics; limited or no access to public and social services and to participation in community-based organizations that are welcoming and affirming; and barriers to housing, banking, and credit alongside inclusion in community services and local elections (Menjívar and Kanstroom 2013). On a daily basis, individuals' immigration statuses may be less or more salient to most of their activities (Coutin 2000: 40). Migrants may be regular in one sense and irregular in another; they may be fully excluded from the legal-political system but able to carry out a range of social interactions and activities.

This observation is important. It draws our attention to the valences of consequences of illegality. Access to various domains of everyday life is shaped by a constellation of entry points and barriers across time, space, and place, offering a variety of configurations of rights and limitations. What's more, where one resides, the absence or availability of resources, the attitudes of local actors and institutions, and individual characteristics such as age, gender, and race shape one's experiences along a continuum from integration to exclusion (Coleman 2012: 181; see also Sigona 2012: 51).

While citizenship is typically imagined as the most desirable condition for residents of a given country and is assumed to mean full and equal participation within society, this is not the case for many individuals who possess formal citizenship. As everyday examples show us, community belonging is never fully experienced, while exclusion is also seldom absolute. One may possess formal citizenship but may experience discrimination due to race, gender, or sexual preferences and, as a result, be positioned on the margins of society. At the same time, a person may lack formal legal status yet be centrally involved in the social and cultural fabric of her or his community. In some locales, undocumented migrants have access to some public services such as education and health care, and in some circumstance may even be permitted to vote in local elections (e.g., Commonwealth subjects in the United Kingdom or undocumented migrants in the United States voting in school board elections).

Legal scholar Linda Bosniak reminds us that the boundary between "legal" and "illegal" is at times sharp but at others fuzzy. It is within the fuzzy spaces that a different perspective has evolved: that which views borders as more permeable and highlights the spectrum of gray spaces between "legal" and "illegal" statuses (Goldring, Berinstein, and Bernhard 2009). This perspective provides a framework through which to assess the factors that mediate the relationship between macro-level structures (global processes and immigration policies) and micro-level practices (the everyday decisions and actions of migrants) and allows us to capture the dynamic processes of contemporary membership. In daily life, these processes shape experiences that fall between the dichotomous experiences of exclusion and belonging. Migrants who have little or no formal rights interact every day with a multitude of state agencies, community institutions, and individuals.

These conditions make it possible for both citizens and migrants to sometimes operate "as if the boundaries did not exist" (Benton 1994: 229; cf. Kubal 2013).

Research exploring the spaces between legality and illegality has demonstrated that migrants' experiences of illegality and belonging are often dependent on specific and situational contexts (Coutin 2011; Gleeson 2009, 2010; Ruhs and Anderson 2010; Rytter 2012). Even when migrants are excluded from the legal-political system, spaces of inclusion determined by local policies or by institutional practices make it possible for individuals without legal status to engage in community activities, establish relationships, and have children. In some locales, they also receive critical services that address everyday needs (e.g., health care and police protection) and are afforded opportunities to more fully participate in community life (e.g., by serving on local councils or voting in local elections) (Jones-Correa 2005; Lewis and Ramakrishnan 2007; Marrow 2009). In some cases, local policies allow undocumented migrants to obtain driver's licenses and rent apartments. In other places, local institutions and/or bureaucracies circumvent the legal system to offer medical care, free legal services (translators and interpreters), and advocacy. Scholars suggest that certain bureaucratic agencies and their staff draw from a professional ethos that sees immigrants as deserving and act to provide them with services as they would to any other clients (Marrow 2009). The response of these bureaucrats – e.g., social workers, healthcare employees, and teachers – allows immigrants important forms of access and enlarges their social worlds.

In addition to experiences that fall outside of dichotomous experiences, there are a growing number of immigration statuses that occupy the gray space between legality and illegality. Scholars generally agree that illegality is a legally sanctioned and routinely produced status and that the social and political conditions under which people are constructed as "illegal" should be critically examined (De Genova 2002). At the same time, however, there is a growing recognition that the state produces different categories of non-legal or less-than-legal statuses. There are many immigrants today who occupy statuses that are temporary, uncertain, and non-linear. Changes in or violations of immigration laws or bureaucratic requirements can move immigrants back and forth between legal and unauthorized statuses (Jasso et al. 2008).

Indeed, the boundaries between legality and illegality are often unclear (Menjívar and Kanstroom 2013). Moving away from the binary categories of "legal"/"illegal," "documented"/"undocumented," several international scholars of migration have argued that traditional, dichotomous ways of viewing immigration statuses are not sufficient (Aleinikoff 1997; Düvell 2008; Goldring, Berinstein, and Bernhard 2009; Kubal 2013; Menjívar 2006; Reeves 2015). Examining the spaces between these dichotomous categories exposes a diverse set of immigrants who – depending on their modes of entry into host countries, their motivations for migrating, the timing of their arrival and, in the case of children, their condition as dependent or independent immigrants – encounter different institutional and policy arrangements. As Anderson and Ruhs (2010) argue, the division of immigrants into mutually exclusive and jointly exhaustive categories – either "legal" or "illegal" – is hardly clear in policy or practice, nor in immigrants' conceptualizations of their status. Tomas Hammar has applied the term "denizens" to those legal residents of a country who are not naturalized citizens but who enjoy some economic and political rights (Hammar 1990; cf. Reed-Denahay and Brettell 2008). Luin Goldring and her colleagues employ "precarious" to describe the multiple pathways to an irregular status among immigrant populations in Canada (Goldring et al. 2009). And Agnieszka Kubal advances "semi-legality" as a term to denote ambiguous relationships with the law (Kubal 2013).

In her examination of Central American immigrants caught in the legal limbo of Temporary Protected Status (TPS), Cecilia Menjívar advanced the concept "liminal legality" to highlight the gray areas in which many immigrants live. While TPS holders are sometimes able to renew permits, a period of non-renewal (due to lengthy processing times or denial) can push someone out of status, even if temporarily. Menjívar observed that this legal limbo can persist indefinitely, not leading to citizenship or other forms of formal legal immigration status. This long-term uncertainty, or "permanent temporariness," is characterized by a sense of ambiguity, moving between legal *and* "illegal" statuses. Additionally, Menjívar highlights the impact of uncertain immigration status on migrants' families and their children's educational prospects. This limbo can also affect children who are citizens but who have undocumented parents (Dreby 2015), as well as citizens with undocumented spouses (Gomberg-Muñoz 2011), who may experience forced separation from their family members due to deportation. This phenomenon hints to a status loss by proximity, with the impact of precarious legal status reverberating horizontally to other documented family members (López 2015) and vertically across generations.

Sociologist Agnieszka Kubal (2013) draws attention to the numerous examples whereby undocumented migrants experience a condition of "semi-legality": that is, an experience of membership that might contain elements of inclusion and exclusion at the same time and that might shift depending on time, space, and place. Examples of semi-legality are abundant: many migrants who today possess a legal status were once undocumented; legal permanent residents may be subject to losing their status and deportation for relatively minor offenses; certain laws may define one's entry as "illegal" but, under certain circumstances, they may also provide legal permission to remain in the country; other laws may define a range of legally permissible weekly work hours for visa holders but may deem anything in excess of those hours illegal.

Possessing forms of legal status does not always guarantee protection against precariousness and vulnerability to state power. Legal residents can also be subjected to deportation. In the United States, for example, a criminal conviction can lead to deportation of lawful permanent residents (Golash-Boza 2016). Indeed, mass deportation due to restrictive changes in US immigration law has given a renewed significance to territorial belonging and sharpened the distinction between citizenship and other forms of legal immigration statuses.

Ultimately, the forms of belonging that exist in the gray spaces between inclusion and exclusion are conditional, partial, temporary, and revocable. Belonging is fragile and contradictory. In some places and contexts, local actors may provide spaces of access and inclusion; in other places and contexts, they may impose additional levels of enforcement, exclusions, and difficulties for migrant populations. At any moment, inclusionary policies and practices can be overturned or curtailed, depending on the willingness of local institutional agents and legislators. It is important to note that even immigrants who have been enjoying spaces of inclusion can be apprehended, detained, and removed from the country.

Conclusion

This chapter examined illegality and citizenship as the two ends of a spectrum that encloses a diverse array of possible in-between statuses. On the one side, illegality is a particular social condition that is regularly and systematically produced and reproduced, as immigration law and labor demand work in tandem to maintain a surplus marginalized and vulnerable workforce. In this sense, the differentiation between and the categorization into legal and unauthorized subjects is primarily marshalled to govern migrant labor through enforced vulnerability and marginalization. On the other side, the chapter discussed the dual and complex nature of citizenship: legal citizenship draws the line between those who have legal permission to reside in and move about a country, while other forms of citizenship based on belonging allow undocumented persons to forge their own paths and make claims to be included. We briefly introduced several terms coined by scholars to describe such alternative forms of citizenship: "global citizenship," "transnational citizenship," "post-national citizenship," "social citizenship," and "multicultural citizenship."

Subsequently, we shifted the focus away from the ways in which illegality and citizenship have been theorized and towards the everyday embodied implications and social processes that

Theorizing the Lived Experience of Migrant Illegality

shape experiences of illegality. Sergio's story illustrated some of these dimensions of undocumented life. While much of his childhood was shaped by his mother's undocumented status, his own undocumented status came into sharp focus during his adolescence. Sergio's lack of status had ripple effects on different aspects of his life and forced him further into the shadows, making his world even smaller. Much like Sergio, undocumented migrants seldom experience their status as an isolated social marker. Rather, their experiences of illegality are shaped by the spaces they traverse and by their different interactions: tight labor markets that are prone to abuse and harsh conditions but where they meet other co-ethnics; limited or no access to public and social services and participation in community-based organizations that are welcoming and affirming; barriers to housing, banking, and credit alongside inclusion in community services and local elections.

In this chapter, we outlined how these dimensions of undocumented life are often contradictory. They fracture relationships, and they entangle undocumented migrants into brutal and unpredictable bureaucratic systems. We subsequently documented how this reality, coupled with the displacement of border and immigration controls from immigration authorities to frontline civil servants or even private citizens such as landlords exacerbates migrant communities' sense of vulnerability. In this sense and through this vantage point, the chapter illustrated that illegality is a dynamic, sociohistorical process, rather than a static concept. It is both produced *and* experienced.

Illegality is typically constructed at the level of the destination country. However, we have argued that it is also important to examine illegality from a global perspective and in the interplay between different scales of governance. This broadened framework adds a conceptual complication: in the absence of the view of the state, what is "illegal?" In other words, if "illegality" is the product of the law and the law stems from a sovereign state, how can we think of undocumented migration as a global phenomenon?

A potential solution to this conceptual conundrum lies in much more practical terrain: that of diplomatic relations between states.

Theorizing the Lived Experience of Migrant Illegality

This idea has been captured in preparatory work and negotiations at the United Nations surrounding the Global Compact for Safe, Orderly and Regular Migration (GCM).

3

Geographies of Undocumented Migration

In 2009, the British right-wing politician Boris Johnson, then mayor of London, called for an "earned amnesty" for undocumented migrants. A study by the London School of Economics had just estimated that there were approximately 725,000 undocumented migrants living in Great Britain, over half of them in London. Explaining the motivation for his liberal position, Johnson proclaimed: "It would take the authorities over 60 years to remove the current number of irregular immigrants on current trends."

It didn't take long for Johnson's party colleague and Home Secretary Theresa May to issue a statement responding to his proposal. "Our policy on an amnesty for illegal immigrants remains unchanged ... those here illegally should go home, not go to the front of the queue for jobs and benefits."[1] Boris Johnson later became foreign secretary in a government led by Prime Minister Theresa May, and along with the change of scale of his political ambition, from local to national, his position on undocumented migrants became less accommodating.

In the United States during this time, President Barack Obama made several attempts to push comprehensive immigration reform through Congress, including a pathway to citizenship for millions of undocumented migrants. While his efforts were met by insurmountable political opposition, some parts of the country were mounting their own push to attract immigrants. Cities devastated by decades of population decline and urban decay, from Baltimore to Philadelphia, actively encouraged immigrants to move in, regardless of their immigration status. As former Baltimore Democratic Mayor, Stephanie Rawlings-Blake, explained, "To get Baltimore growing again, it makes sense to look at what made us grow in the first place: the strength of our immigrants."[2] Seeking to attract 10,000 new families to Baltimore over the span of a decade, her hope was that the newcomers would boost income and property tax revenues and create jobs, thereby reinvigorating the city's economy. To show that she was not differentiating between types of immigrants, in 2012 Ms Rawlings-Blake signed an executive order prohibiting any city police officer from asking residents about their immigration status.

In the previous chapters, we have introduced the main concepts and theories on undocumented migration and have offered a glimpse into the experiences of illegality. In this chapter, we shift the focus to the subnational and local levels. Specifically, we examine how different cities and other locales in Europe and in the United States have addressed the presence of sizeable undocumented migrant populations. Additionally, we explore the part that local contexts play in producing experiences of belonging and exclusion and in shaping the emergence of new political subjectivities and voices. We also observe how borders operate internally, within cities and neighborhoods, and how these "bordering" practices shape the perceptions and experiences of undocumented residents.

Here we locate undocumented migration at the subnational and local scale and consider how a shift of perspectives enables a different aspect of the phenomenon to emerge. Drawing on a series of case studies, we highlight the dynamic dimensions of more local geographies. In the United States in the absence of comprehensive federal solutions to immigration, these subnational contexts can be spaces where undocumented migrants can articulate a meaningful presence and assert claims for recognition and belonging. However, these locales are not exempt from the pervasive presence of immigration control and enforcement. Borders have become increasingly present inside the territory of the state, incorporated within public services, privatized by for-profit companies, transferred to individual landlords who are forced to check prospective tenants' visas, or bestowed on the public (Flynn 2005).

Globalization and its local manifestations

Urban areas have long attracted internal and international migrants. As a result, cities have become increasingly diverse and transnationally connected. Since the mid-2000s, the majority of the world population lives in urban areas. Moreover, a sizeable number of people are concentrated in a relatively small number of megacities, like New York, Tokyo, and Delhi. According to UN estimates, 9 percent of the world population will be concentrated in 41 megacities by 2030. According to sociologist Saskia Sassen (1991, 2000), this process of simultaneous migration and urbanization is the product of the restructuring of global capitalism that necessitates the concentration of capital, resources, and infrastructures in a series of strategic global cities that make capital hypermobility possible. These cities are crucial for the dual global flow of information and capital. The spatial dispersal, digitalization, and deterritorialization of economic activities associated with globalization depend on the territorialization of infrastructure and labor that makes it all possible.

These processes exacerbate inequalities as they increasingly marginalize large swaths of the urban population. As Sassen points out, "If we consider that global cities concentrate both the leading sectors of global capital and a growing share of disadvantaged populations ... then we can see that cities have become a strategic terrain for a whole series of conflicts and contradictions" (Sassen 1991: 39). Migrants are at the core of these tensions. Sassen's vantage point allows us to reframe international migration in terms of the internalization of the labor market, as "a set of processes whereby global processes are *localized,* international labor markets are constituted, and cultures from all over the world are de- and re-territorialized" (Sassen 2000: 83). As capitalism has reached the global stage, cities have become the ideal location for these processes to thrive. Urbanization attracts both internal and international immigrants to fill jobs at various rungs of the employment ladder.

Migration has been largely studied through the lens of the nation-state, which has had profound implications for the ways in which immigrants are perceived, portrayed, and treated. Thus immigrants are often constructed as threats to the security, identity, and culture of receiving countries. Cities, however, provide a different vantage point from which to observe the impact of migration on society. As such, the urban scale enables us to cast light on the complex nexus between the local and the global brought about by neoliberal restructuring, as well as to observe articulations of membership and belonging which challenge, with varying degrees of success, the state-centered politics of belonging and construct a local, more inclusive, imagined community.

Immigrants are integral to cities. They are more than just residents in an urban landscape, they form and transform urban environments: "as part of the labor force upon which cities build their competitiveness; as historical agents; as agents of neoliberal restructuring who contribute or contest the changing status and positioning of neighborhoods and cities" (Schiller and Çağlar 2009: 12). In particular, immigrants in cities are active agents in the rescaling processes described earlier, rather than just a consequence of them.

Local governments play an important role in matters of immigrant integration and in crafting policies and initiatives to address immigration issues in their locales. In their efforts to address immigrant issues in their own locales, local lawmakers often come up with their own responses that supplement, buffer, or rebuke national immigration policy. These actions may take place at various scales – community, municipality, county, province, or state – and across a range of authorities (Varsanyi 2010). Some governments and local authorities adopt restrictive policies against migrants, while others promote broader inclusion.

Both in the United States and in Europe, differences in state and local laws, policies, and approaches to law enforcement abound. In the United States in particular, some local law enforcement units have signed agreements with federal immigration agents, while others have refused to cooperate with federal authorities in matters of immigration enforcement. Both in the United States and in the European Union, some cities have designated themselves as "sanctuary cities" and work to shield immigrants from immigration enforcement while others offer no such protection (Ridgley 2008). These contradictory processes create an uneven patchwork of legal conditions that powerfully shapes the everyday experiences of undocumented migrants and their families.

Institutions and community organizations also play a crucial role in these matters. Many communities within urban areas have well-developed social services and educational infrastructures that offer programming and services in immigrants' native languages. Local organizations often provide a range of services, from legal assistance to mental-health support. They assist with everyday needs and help migrants build language and workforce skills. They also offer immigrants opportunities to build social capital and engage in civic opportunities. And they can provide spaces and activities where immigration status is not consequential to participation. However, many immigrants today live outside of urban spaces and do not have access to many of the services and resources mentioned. And when local police, employers, and other public services become an extension of immigration enforcement, community organizations can reproduce vulnerability and sow the seeds of fear and anxiety.

Indeed, locales can exacerbate the stress of living without full legal status, while they can also foster community participation and belonging. As the case studies from New York and Barcelona later in the chapter show, cities are, at least to some extent, independent actors able to define the boundaries of urban citizenry and their own politics of belonging. These cases compel us to reconsider debates on integration and inclusion beyond the national framing, from a local standpoint. But the local level is also a terrain where international, national, and local agendas meet and sometime clash. Areas with high concentrations of undocumented migrants are often targeted by heavy federal immigration enforcement activity. Local public services become sites for immigration control, and landlords and employers are under pressure to perform immigration duties.

Everyday bordering

"The aim is to create here in Britain a really hostile environment for illegal migration."
Theresa May, May 25, 2012

In matters of immigration, the mismatch between declared policy objectives and policy outcomes has long attracted the attention of migration scholars (Boswell 2007; Castles 2004; Joppke 1998; Lahav and Guiraudon 2006; Sassen 1996; Soysal 1994). In particular, scholars have observed the gap between the tough talk on migration that often frames immigration debates and the sometimes more liberal outcomes regarding the admission of foreigners and their access to various social and economic rights (Martin, Cornelius and Hollifield 1994). In fact, while a lack of immigration status leads to an exclusion from a range of rights and entitlements, in everyday practice this exclusion is almost never absolute but usually intertwined with simultaneous processes of partial inclusion (Chauvin and Garcés-Mascareñas 2012; Cvajner and Sciortino 2010; Sigona and Hughes 2012). According to Boswell, one of the reasons for this discrepancy is the involvement in immigration enforcement of a greater number of institutional actors who do not necessarily share the priorities of immigration authorities (Boswell 2007).

Wealthy and more populous cities, like New York City and London, can achieve greater autonomy from national governments and craft their own policies of varying levels of inclusion and exclusion. At the same time, newer geographies of migrant settlement struggle to come to terms with diverse populations of newcomers while also learning how illegality plays a role in their settlement. Cecilia Menjívar (2016) suggests that new spaces of irregularity are produced in the interstices between federal and state legislations in the United States. It is in these spaces that the boundaries between legality and illegality become increasingly permeable. For undocumented migrants, these spaces pose challenges to their understanding and navigation of the legal ambiguities that are created as they try to construct meaningful lives. Often, this endeavor is complicated by the distinct and even conflicting regimes of rights and visions of memberships that can coexist in certain geographies.

In 2012, the UK government set up the Allegations Management System to record intelligence leads on irregular migration from the public. A few months earlier, then Prime Minister David Cameron announced the intention to crack down on unauthorized immigration, urging the public to help reclaim the British borders and to report those suspected of being undocumented migrants. According to a report by the UK Independent Inspector of Borders and Immigration, Cameron's invitation led to thousands of allegations (ICIBI 2015). In the twelve months between August 2014 and July 2015, a total of 74,617 allegations were lodged on the system, with more than 49,000 coming directly from the public via the Immigration Enforcement hotline, electronically through the Gov.uk website, or in person to officers (CorporateWatch 2016). Noteworthy, less than 2 percent of tip-offs led to the arrest and removal of undocumented migrants (Home Affairs Select Committee 2013).

While not an efficient tool of immigration enforcement, the Allegations Management System was effective in two important ways: namely, it simultaneously provided the public with a sense of protection while making a person's undocumented status apparent and the center of the problem, thus reinforcing the idea that people are "illegal" and threats to security. It also exacerbated public anxiety on immigration, contributing to the production of a hostile environment towards undocumented migrants or anyone deemed to be undocumented (often individuals from racialized minorities, including citizens).

In Europe, the lifting of passport control and border checks between several EU member states in the Schengen area[3] has led to a diffusion of immigration controls in hospitals and other public institutions and spaces. In Athens, Greece, for example, between 2012 and 2015, the police operation Xenios Zeus targeted "migrant-looking" individuals through racial profiling in public spaces across the city center, leading to the detention of more than 80,000 people (Dalakoglou 2013). Similarly, the Swedish REVA project, launched in 2009 by the Swedish Prison and Probation Service (Kriminalvården) and the Swedish Migration Board (Migrationsverket), racially profiled members of certain groups in the public transportation system (Keshavarz and Zetterlund 2013). Finally, the EU border force's (Frontex) operations Aerodromos, Aphrodite, Perkunas, and Mos Maiorum (Heller and Jones 2014) explicitly and routinely target, apprehend, and investigate migrants inside EU territory. These efforts reveal the ways in which undocumented migrants are racialized, further "othering" their existence within the state.

The vanishing borders within most of the EU territory has meant that, increasingly, all intra-European public and transit spaces have become strategic places for border-control enforcement (Amoore, Marmura, and Salter 2008; Bigo 2001; Coleman 2007). Evoking the threat of terrorism, as well as the construction of migration as an invasion, this internal discourse (Bigo 2001) serves to redeploy the border in different ways inside EU territory – in town squares, train and bus stations, schools, and workplaces.

This process has also meant that we can no longer think of borders in the terms of solid infrastructures demarcating state borders. Rather, border control now occurs routinely and diffusively in everyday life. Discussing the position of EU academics in the United Kingdom at a time in which their immigration status is called into question in the Brexit process, Nando Sigona (2018) writes:

> Thousands of non-UK citizens working in UK universities, including many EU nationals, find themselves in a paradoxical position: at the same time co-opted in the job of border guarding their students, for example collecting signatures, reporting unjustified absences and even being granted the power to decide if an international student can or cannot travel back home for a wedding or a funeral – while increas-

ingly experiencing themselves the "hostile environment": border guards and border guardees.

Veiling harsher enforcement and security measures under the rhetoric of public safety and anti-terrorism initiatives places undocumented migrants and their families in even more vulnerable positions. As the US–Mexico border has become increasingly dangerous and deadly to cross, its reach has also been extended beyond its geographical bounds. In effect, crossing the border into the United States does not guarantee a long stay nor a more clandestine life within the confines of the state. As some cities and municipalities have signed 287(g) agreements and enforced harsher employment and restrictive higher education policies, undocumented families face some of the toughest challenges from everyday bordering.

The following case studies from New York, South Carolina, Barcelona, and Athens offer local insights into some of the processes of bordering and membership discussed in the preceding pages. They point to the emergence of localized forms of membership and belonging as well as to the far-reaching hand of immigration enforcement within the territory of the state. They also highlight the tension between different visions of membership as they are articulated at the local scale and in everyday encounters.

"A distinctly New York story"

In his 2015 State of the City speech, New York City Mayor Bill de Blasio spoke of the struggles and triumphs of his Italian grandmother who migrated to New York as a young woman. This rags-to-riches story was a familiar one to millions of New Yorkers, celebrated in songs, films, museums and theaters, and circulated all over the world. But de Blasio's story was not merely a personal one. The story serves as a reminder that, in a city like New York, the overriding theme of perseverance resonates with many of its residents, even if only as an aspiration. New York is the city that created and nurtured the unique conditions that enabled "the tired, the poor, and the huddled masses," celebrated on the Statue of Liberty, to build the wealthiest democracy in the world.

Expressing frustration over the federal administration's stalemate over comprehensive immigration reform, De Blasio's evocation of the city's history of migration and diversity served to frame a new initiative launched by the city government – the New York City Identity Card (IDNYC) that offers the city's estimated half a million undocumented migrants a municipal identification card and "secures the peace of mind and access to City services that come from having recognized identification."[4] Through this highly symbolic initiative, the city government stakes a claim to define who belongs to the city, regardless of their position vis-à-vis the federal government. It does so in an inclusive way, not specifically targeting New York City's undocumented residents but creating for the eight million New Yorkers living in the vast, diverse, and divided city an "imagined community" where they can all belong (Anderson 2006).

As of January 2015, all residents of New York over the age of fourteen, regardless of immigration status or criminal record, were eligible to apply for an IDNYC card. By brokering access to services and businesses with which city residents are likely to interact, these cards offer a creative way to address some of the challenges posed by federal immigration law for undocumented migrants.

> By treating undocumented status as merely one of many factors inhibiting a New Yorker from being able to fully engage with the city's services and institutions, IDNYC challenges the traditional role cities play in the citizenship of their residents . . . The multidimensionality of citizenship hints that actors other than the federal government can and do play a legally permissible role in citizenship enhancement" (Torres 2017: 338, see also De Graauw 2016).

The establishment of IDNYC continues the city's legacy of being a hub for immigrants. The city has served as a continuous gateway for immigrants from all over the world (Waters and Kasinitz 2013). Its long history of immigration has allowed New York to adapt and evolve with every wave of migrants that reach its shores. Given the city's growing infrastructure, expansive transportation system, and unparalleled diversity, it is not surprising that undocumented migrants continue to find and craft a home for themselves there. Ethnic enclaves have thrived, smoothing the transition for newcomers while also providing them with work and community opportunities. In addition to its long-standing immigrant history and ethnic enclaves, New York City also boasts a wealth of institutions and community organizations created specifically for the needs of immigrants, regardless of immigration status (De Graauw 2016). Government offices, such as the Mayor's Office of Immigrant Affairs, attempt to portray a sense of inclusion for immigrants and have "pledged to make New York City a safe and open community for all who live here, regardless of immigration status."[5] These forms of inclusion provide undocumented migrants with formal avenues to necessary resources and information. Through these networks and institutional forms of support, undocumented migrants can live less clandestine lives and access modest forms of mobility. This is especially true for undocumented students who have been granted in-state tuition at the state public institutions of higher education.

However, not all cities have the same power to create their own model of inclusive citizenship, and their room for maneuvering is subject to continuous negotiations with state and federal authorities. The US government's recent threat to withdraw federal funding from cities that provide certain protections to undocumented migrants and refuse to cooperate with federal immigration officials (so called "sanctuary cities") has illustrated the vulnerability of some local governments working to define inclusive citizenship for themselves.[6]

What's more, as much as cities like New York can be commended for their positive contexts of reception, many of their immigrant residents struggle to make ends meet. Soaring housing markets and higher-than-average living costs have greatly impacted the everyday lives of their residents. Gentrification and other forms of urban renewal have had considerable negative effects on the poor and working class. Through unprecedented rent increases, building buy-outs by development companies, and the decrease in rent-controlled units, long-time residents have been displaced from their homes. For undocumented migrants, among the city's most vulnerable inhabitants, a lack of access to high-paying employment coupled, with the dearth of affordable housing options, undercuts efforts to experience membership and community belonging.

South Carolina: the devolution of immigration policy and negative state contexts

Over the last two decades, the United States has experienced a significant demographic shift of Latin American migration from urban areas in traditional immigrant gateway cities to more rural destinations in the South and Midwest (Marrow 2011; Massey 2008; Zúñiga and Hernández-León 2005; Zúñiga et al. 2002). Significantly, one-third of recent Mexican immigrants have settled outside of traditional gateway cities, and more than one in every five lives in rural towns (Lichter et al. 2010; Singer 2004; Stein et al. 2016). Scholars studying these new and re-emerging growth areas cite ready access to work and lower costs of living as key factors in positive economic incorporation for immigrants. In addition, the rise in agricultural labor, food-processing factory work, and other forms of employment have attracted immigrants from US cities (Light 2006) and have prompted them to migrate directly to these destinations. However, many of these locales lack adequate social service infrastructures for immigrant families and provide only limited opportunities for their children (Allard and Roth 2010).

The state of South Carolina is one of these new destinations. The last three decades have seen an increase in the overall population growth in states like South Carolina, with immigrants driving sizeable shares of that growth. As a result, the state's changing demographic landscape has brought about cultural, social, and political changes.

Due to political gridlock in Congress over the last several years, the inability to overhaul federal immigration legislation has prompted several states to take matters into their own hands, resulting in a proliferation of immigration proposals at the local level. While certain states have broadened access to the polity, offering undocumented migrants the opportunity to apply for driver's licenses and to receive in-state tuition at public universities, others have adopted a more restrictive stance, attempting to criminalize unauthorized presence and exclude undocumented migrants from public universities. This uneven geography of local enforcement ties where one lives to a multitude of experiences based on local impediments and opportunities.

As such, undocumented migrants in South Carolina face a host of challenges. While migrants living in the state might be able to find multiple avenues to employment and experience a lower cost of living relative to urban cities like New York, they experience a multitude of challenges accessing transportation options and finding immigrant-friendly organizations and services in their language (Roth, Gonzales, and Lesniewski 2015).

According to the US Department of Agriculture, rural America makes up about 16 percent of the total population of the United States. Yet nearly 47 percent of South Carolina's population resides in rural or largely rural communities. These conditions create challenges for many of its newcomer population, particularly those who lack legal immigration status and do not speak English (South Carolina Community Loan Fund 2013).

Latin American-origin immigrants in new rural destinations like South Carolina encounter a host of issues that negatively impact their social and material well-being: concentrated poverty; a lack of stability in employment; a dearth of social service organizations; and a lack of public transportation options (Gonzales and Ruiz 2014). While poverty rates tend to be higher in rural areas, Latin American immigrants experience acute poverty at higher levels than any other ethnic or racial group (Kandel et al. 2011; Lichter and Johnson 2007). And while rural America may provide an easier entrée into the low-skilled workforce, its low wages and the transient nature of agricultural jobs limit opportunities for upward mobility. In addition, rural communities' weak social service infrastructures complicate immigrants' abilities to receive critical services, such as ESL classes and workforce training, that would better facilitate their integration into the larger community (Donato et al. 2007; Pruitt 2007). Also, limited public transportation options can be particularly troubling for undocumented migrants in states that do not allow them to obtain driver's licenses.

The settlement of undocumented migrants in South Carolina also exposes the racialized aspect of illegality. Latin American migrants, while segregated and socially isolated, are also disproportionately profiled and targeted by law enforcement officials. South Carolina is one of 38 states that does not provide driver's licenses to undocumented migrants, making it harder for undocumented migrants to navigate rural life. Furthermore, four counties in South Carolina have partnered with US Immigration and Customs Enforcement (ICE) through 287(g) agreements. Named for the section of the amended Immigration and Nationality Act that authorized such memoranda of agreement, 287(g) agreements provide training in local law enforcement and delegate authority to local law enforcement to enforce immigration law within its jurisdiction (Gonzales and Raphael 2017). This means that local law enforcement works with ICE "in order to receive delegated authority for immigration enforcement within their jurisdictions" (ICE 2018a). Thus, for an undocumented migrant in South Carolina driving to work, a simple traffic stop by local law enforcement can lead to arrest, detention, and deportation. Over the last decade, hundreds of thousands of immigrants across the United States have been placed in removal proceedings after being arrested or cited for minor traffic violations (Alonzo et al. 2011). These practices have had the effect of ratcheting up fear and anxiety in local communities.

Another realm in which undocumented migrants in the state face difficulty is through higher education. South Carolina is one of three southern states that deny access to public higher education for undocumented migrants. Undocumented students are unable to attend public universities and colleges in South Carolina and are ineligible for in-state tuition even if they could. Thus undocumented migrant youth who have attended public K-12 schools in the state are forced into the same labor markets as their parents, further increasing their vulnerability.

As we have delineated with the earlier example of New York

City, some locales have both governmental and nongovernmental infrastructures to provide support and mitigate some of the burden of illegality for undocumented migrants. Cities and towns in South Carolina, on the other hand, rely primarily on nongovernmental organizations and social institutions to do the work of incorporating new arrivals and undocumented migrants. Churches, local public schools, and local non-profit organizations have taken on much of this work in the south (Marrow 2011; Winders 2012). These informal and formal forms of support allow undocumented migrants to survive in areas that are not familiar to them and are at times hostile towards them. Yet many of these organizations struggle to fund their programs and cannot adequately address the needs of migrants and their families. What's more, many of these important services are spread across the state, requiring long commutes.

Barcelona: migrants and the local politics of belonging

During the month of February in 2017, some 160,000 people demonstrated in the center of Barcelona to demand that the government allow more refugees into Spain from war-torn areas like Syria. Marchers carrying placards and banners – many in the Catalan language – accused the Spanish government of dragging its feet over the question of refugee resettlement, having not honored its pledge made in 2015 to allow more than 17,000 refugees into Spain within two years. Over that time, Spain only accepted about 1,100 refugees.[7]

Barcelona, the capital city of the Spanish autonomous region of Catalonia and Spain's second-largest city, has a long-standing reputation as a welcoming and inclusive city. In recent years, the city has received international praise for its initiatives vis-à-vis resident migrant communities and newcomers. Barcelona positioned itself as a sanctuary city, and other cities have looked to Barcelona as a model of virtuous interaction with migrant communities. This position was confirmed during Europe's "refugee crisis" when the city administration took a leadership role among European and international cities to offer shelter and hospitality to refugees (Bauder and Gonzalez 2018).

Following the mayoral election of Ada Colau, an independent left-wing politician who became known for her involvement in the anti-eviction campaign *Plataforma de Afectados por la Hipoteca* in Catalonia, the city of Barcelona invested in the creation of services at the municipal level to curb anti-immigration national policy. Under Colau, the Barcelona City Council has promoted an image of the city as welcoming to migrants.

Similar to New York City's Identity Card program, Barcelona's *documento de vecindad para inmigrantes* allows undocumented migrants to apply for a municipal document asserting their membership of the city by virtue of their residency (Garcés-Mascareñas 2018). Importantly, the document is meant to help shield irregular migrants from detention and potential deportation. In the face of the growing population of undocumented migrants in the city and the increasingly difficult pathways to regularization and citizenship, the city government launched the local ID card initiative as an alternative ID card that would act as a de facto regularization of undocumented migrants. However, such decisions still rest largely on the discretion of the courts.

Another policy that is seen to have ameliorated the conditions under which undocumented migrants live in the city is the changing of the city's rules for *empadronamiento* (registration of residency) (Bermúdez and Escrivá 2016). The city administration reformed the existing rules in order to make it possible for a person's usual residence to be recognized as his or her legal residence, even in the absence of a contract. Since this is usually the most common bureaucratic barrier for undocumented migrants, in this way, through *empadronamiento*, many have now gained access to public services including health care. Taken together, these initiatives are indicative of the city administration's effort to better understand the conditions of its migrant population and to be responsive towards the city's grassroots social movements.

However, for the young migrants, usually from sub-Saharan Africa, trying to make a living out of street-vending goods, none of these really does much to shield them from police abuse and violence. The complicated issue of the *manteros* (street vendors) offers a competing perspective on the local politics of migration, its idiosyncrasies and contradictions. The rapid increase in tourism in Barcelona over the last few years has led to an increase in street vending, an activity the city government and shop owners view negatively due to its unregulated nature and public space issues (Moffette 2018). In efforts to resolve these problems, along with campaigns to inform tourists not to buy from unlicensed street vendors, police repression has also increased, including targeted raids and roundups of street vendors, very often resulting in scuffles. At the same time, the City Hall has issued a directive prohibiting any police intervention in the spaces inside the metro and railway stations, creating de facto safe spaces for *manteros*, yet ones that are precarious and subject to changing political will.

Athens: the city as a space for political mobilization

Following the refugee crisis of 2015–16, Athens has attracted a sizeable population of forced migrants, many with uncertain or pending legal status. Their settlement in the city has occurred at a time when an economic and financial crisis has dramatically downsized budgets of local authorities leaving refugee camps strapped for resources. Moreover, in Greece, there is little room for local authorities to contribute to policy making in this matter, as decisions about funding rest in the hands of the national government and the European Commission.

Athens has seen a considerable increase in its migrant population since the refugee crisis in 2015 and 2016. In particular, following the European Union's refugee agreement with Turkey[8] and the closure of the Greek–Macedonian border and the Balkan route in March 2016 (Crawley et al. 2017; European Council 2016), thousands of migrants and refugees were trapped in Greek territory, many of them still stranded in Greece awaiting a decision about their asylum cases and applications for resettlement (Papoutsi et al. 2018). While Greece does not typically feature high on the list of preferred migrant destinations, due to more than ten years of economic and financial crises and austerity policies, the European Union's policy of migration containment has turned the country into its buffer zone, placing a strain on Greece's social infrastructure and limited resources. Athens, Greece's capital, is no exception. As the city has received the bulk of stranded asylum seekers, its understaffed schools have exceeded capacity, struggling to manage multicultural and multilingual classrooms. Meanwhile, the city's hospitals are pressed to manage growing numbers of patients who do not speak the language and who often suffer from mental health issues such as post-traumatic stress disorder and chronic depression.

Yet the four humanitarian camps that have sprung up around the city to accommodate recently arrived asylum seekers whose applications are being processed cannot provide for everyone. In addition, migrants who have not properly registered upon arrival to Greece (i.e., those moving without proper authorization from one of Greece's frontline islands to the mainland before their application was processed in an effort to avoid the fast-track border procedure) are denied accommodation, access to basic public services, and the scarce allowance afforded to asylum seekers (Vradis et al. 2018).

There is, however, a significant grassroots mobilization in Athens and a convergence of international activists and NGOs in the city trying to meet the needs of the newcomers and struggle with them for their rights. Drawing on these grassroots solidarity initiatives, in 2016 Giorgos Kaminis, the mayor of Athens, proposed the Solidarity Cities initiative.[9] Launched in the framework of the EUROCITIES network, the initiative was created to establish a framework under which European cities could improve access to funding, share information about actions and initiatives, and work together to support efforts to receive and integrate refugees.

However, since 2010 the city administration has also championed a rather exclusionary set of discourses and practices regarding the use of urban spaces by migrants. Additionally, largely driven by the economic crisis, many of Athens's residents (including civil servants) have voiced concerns about capacity issues to support initiatives that would ameliorate the everyday lives of migrants in the city. Most funds are tied up in bureaucratic structures that are managed by the central government to which local authorities do not have easy access. The same is true for many of the NGOs that are active in the city: their highly bureaucratized and professionalized structures are often unable to offer the promised services in a timely manner.

As a result, local authorities and organizations have resorted to grassroots strategies or other indirect means to achieve hands-on interventions in the city. Beginning in 2016, an assemblage of local and international activists has converged on the city, creating a hub of solidarity and sharing knowledge and best practices – an unofficial sanctuary space within the city. A network of eighteen illegally inhabited, or squatted, buildings around the center of Athens houses between 1,500 and 2,000 individuals, while several activist projects provide much-needed services, ranging from medical care to theater and art workshops (Tsavdaroglou 2018). This informal infrastructure has created an alternative support system for thousands of people who either have no access to the official refugee camps and services or, for various reasons, choose not to participate. This network also empowers and politicizes this population, creating an example in direct opposition to the state's practice of segregation and containment in camps outside the city.

Conclusion

This chapter located the study of undocumented migration at the subnational level and offered a view from below – largely from the city. This shift in perspectives enables the analysis of different aspects of the phenomenon. We examined how different cities and locales in the European Union and in the United States have addressed the presence of sizeable undocumented migrant populations. Additionally, we explored the ways in which local contexts help produce experiences of belonging and exclusion, and even shape the emergence of new political subjectivities and voices. We also illustrated how borders often operate internally, within cities and neighborhoods, and how these border practices shape the perceptions and experiences of undocumented residents. This chapter has shown that cities can offer a space for undocumented migrants to mobilize, build solidarity, and claim belonging. Ultimately, the city scale enables the articulation of local understandings of citizenship, yet these understandings contrast at times with the national politics of belonging.

As cities, especially global cities like London and New York, are at the core of globalization processes and neoliberal restructuring, this chapter urges the reframing of undocumented migration within these contexts. Even though migration has traditionally been examined through the lens of the nation-state, it is in cities and particular locales that migrants live, build relationships, go to work, get married, and drive their children to school. This more local vantage point casts light on the impact of migration on society and communities, but also on how everyday bordering and the dislocation of border controls into everyday settings impacts migrants' sense of belonging. The case studies briefly presented in this chapter – New York, South Carolina, Barcelona, and Athens – show that locales can either exacerbate the stress of living without full legal status or can foster community participation and belonging. We illustrated the tensions that oftentimes exist between the different scales of policy making and enforcement and the spaces that may be carved out. These spaces are where the boundaries between legality and illegality are particularly and increasingly permeable.

In particular, our brief case studies detailed local approaches and practices of citizenship that connect undocumented migrants to broader visions of inclusion in the urban realm. As our case studies show, cities often have the power to mitigate the effects of national hostile policies. As the cases of the IDNYC in New York and the *documento de vecindad para inmigrantes* in Barcelona show, cities can effectively shelter people from the impact of their undocumented status. In the case of the United States (and also in more decentralized states, like Germany and Italy), there is a further articulation to consider, namely the relationship between federal and state governments. We saw this playing a central role under the Obama administration, where some states, such as Alabama and Arizona, were taking a more restrictionist stance than the federal government. Or under President Trump, where some states are resisting some of the administration's anti-immigration measures.

Urban space is also a stage for immigration enforcement. Measures that may challenge local understandings of citizenship can range from raids by border police in restaurants and the posting of billboards inviting undocumented migrants to self-report their status, to the refusal of accident and emergency departments to provide help to undocumented patients in need of care. Therefore, the chapter also elaborated on the concept of everyday bordering. The border is redeployed in different ways inside a nation-state's territory, and border control responsibilities have devolved from immigration authorities to civil servants and even citizens.

4

Immigration Enforcement, Detention, and Deportation

"Even as we are a nation of immigrants, we're also a nation of laws. Undocumented workers broke our immigration laws, and I believe that they must be held accountable – especially those who may be dangerous. That's why, over the past six years, deportations of criminals are up 80 percent. And that's why we're going to keep focusing enforcement resources on actual threats to our security. Felons, not families. Criminals, not children. Gang members, not a mom who's working hard to provide for her kids. We'll prioritize, just like law enforcement does every day."
    Remarks from President Barack Obama, November 20, 2014

In early April 2018, a federal ICE raid at a meatpacking plant in the small town of Morristown, Tennessee resulted in the apprehension of 97 migrant workers. This small agricultural town has drawn migrant workers from Latin American countries since the early 1990s when they began working in the region's tomato farms. Over the years, their families have become established in the community. They've enrolled their children in local schools. And they've become active members of their churches and other local organizations. The impact of the raid was felt throughout the town. Employers saw their workforce depleted. Townspeople felt the emotional impact of what many experienced as a traumatic event. Teachers carried out the difficult work of consoling children whose parents had been taken away – as many as 160 children had a parent who was picked up. And spouses, siblings, and older children were left to pick up the emotional and economic slack. Many families relied on the generosity of neighbors and local support agencies. But everybody knew this kind of support would not be sustainable.

The immigration raid in Morristown took place in the wake of several other workplace actions that year, including employee audits at 98 7–Eleven stores in 17 states, including California, Maryland, and Michigan.[1] These efforts have expanded in recent years, due in large part to growing levels of integration between immigration authorities and local police and other public and private agencies (Gonzales and Raphael 2017).

Undocumented migration has risen to the top of the political agenda in most western states. While those immigrants legally residing in receiving nations make up the vast majority of migrants worldwide, the political debate, as well as the policy response, has been driven by a compulsion to expel undocumented migrants (Jones-Correa and de Graauw 2013). Increasingly, heavy enforcement measures – both an increase in border control efforts and a sustained effort to police the national territory – have become the primary mode of addressing undocumented migration. These efforts have enlarged enforceable territory and spread out responsibility and jurisdiction to a larger number of actors and entities (Coutin 2000). As a result, many western nations have seen substantial spikes in apprehensions, detentions, and removals.

Contemporary enforcement measures that extend immigration enforcement from the border to the interior of the country – to small towns, neighborhoods, places of work, and public spaces – have impacted a large number of people, both migrants and those connected to them. They have depleted workforces, disrupted community harmony, and separated families.

These actions have also sown fear and anxiety across communities and families, underscoring the observation that everyday acts like taking a walk in the park, waiting at a bus station, dropping children off at school, and visiting a courthouse could lead to arrest, detention, and deportation. As anthropologist Nicholas de Genova (2002) points out, the ever-present fear of being removed keeps migrants and their families scared and compels them to self-monitor their actions and involvement in community life. As such, the palpable sense of fear they experience draws a tight circle around everyday life. For those migrants belonging to families, enforcement activity can have particularly detrimental effects on family members – even those who possess forms of legal immigration status (Bean et al. 2011; López 2015; Yoshikawa and Kalil 2011). Immigration enforcement, while sometimes a necessary security measure, separates families, forces spouses of deported loved ones into single parenting, produces secondary waves of dislocation, and leaves employers, church congregations and neighborhoods reeling with the loss of valued and cherished members. In this chapter, we examine the practices of contemporary immigration enforcement, its varied forms, and the modern-day implications of heavy enforcement. Initially, we focus on what we call "soft enforcement," the measures put in place by governments to produce hostile environments for undocumented migrants which are expected to act as deterrents for aspiring new immigrants and to lead those already in the country to make voluntary returns to their countries of origin. We successively focus on the detention–deportation nexus and the for-profit industry that has grown around it.

Creating a hostile environment

As we mentioned in chapter 2, sociologists Cecilia Menjívar and Leisy Abrego argue that increased local activity associated with recent immigration enforcement efforts inflict "legal violence" upon individual migrants and their families (Menjívar and Abrego 2012). Accordingly, the criminalization of immigrants operates beyond exclusion; it also generates widespread violence that permeates family life, community activity, schooling, and work, while also dramatically limiting migrants' long-term trajectories.

While detention and deportation represent more visible and overt practices of immigration enforcement, countries also leverage less tangible, but arguably more pervasive, tools that assist in achieving broader enforcement goals. "Soft enforcement" measures, or the practices used to facilitate self-policing and compliance, have widened enforcement efforts and dramatically increased their effectiveness (Carroll 1988; Charles 1982). Immigration checkpoints in neighborhoods and highway on-ramps, local ordinances that prohibit street vending, and policies that compel service providers to ascertain patients' immigration status are examples of such measures that contribute to hostile environments for immigrants. These local manifestations of enforcement achieve broader aims by erecting barriers to employment, help seeking, and mundane everyday pursuits. While these tactics are often used as deterrents for migration, they also operate to keep migrants vulnerable and to press them into considering voluntarily departures, or what former US presidential candidate Mitt Romney once referred to as "self-deportation" (Allsopp, Chase, and Mitchell 2015; Bloch and Schuster 2005).

Often enacted through formal policy, soft enforcement measures operate in a diffuse fashion by erecting barriers to everyday mundane activities and imposing immigration tasks on public service providers, landlords, employers, and even private residents. In this sense, the work of policing borders and enforcing immigration laws has been extended to a range of actors charged with enforcing access to employment, housing, social services, roadways, and public and private space. These strategies have transformed identification checks from one-off encounters between border agents and migrants within the geographical space of national borders into routine and repeated micro-encounters within neighborhoods and public spaces (Sigona 2018). Here enforcement encroaches on everyday interactions in migrants' daily lives and, in doing so, imposes new meaning on them. Pregnant women avoiding interactions with their general practitioners or cutting short hospital stays after delivery for fear of being reported to the authorities for their immigration status is just one implication of these measures (Sigona and Hughes 2012).

The proliferation of these kinds of internal and everyday borders often occurs incrementally, one little step at a time. Typically, these changes are initially challenged. But they often return, slightly repackaged. For example, when in 2018 the UK government introduced measures to impose visa checks on hospital patients, the British Medical Association resisted, arguing that patient confidentiality was the "cornerstone of doctor/patient relationship," and that doctors were not immigration agents (Cooper 2018).

However, the National Health Services (NHS) involvement in immigration checks had started several years earlier. In 2014, the Immigration Act increased restrictions on entitlement to NHS care for foreign nationals. In 2017, legislation in England introduced a mandatory payment system before treatment for those unable to prove their eligibility (i.e., proof of residency status), and a denial of care to those unable to pay. This policy applied to hospitals and a range of community services. It also outlined treatment which is "urgent or immediately necessary" to be provided and possibly charged retrospectively. Importantly, in England the details of patients with unpaid NHS debts above £500 are referred to the Home Office after two months, which can lead to an immigration or asylum application being denied.

Serhiban has lived in London for nine years with her daughter.[2] She is Kurdish from Turkey. Her asylum application was rejected, and her appeal was unsuccessful. For three years, as is often demanded of refused asylum applicants pending forced removal, she checked in with authorities on a weekly basis. Then she stopped. She could not bear the weekly fear and anxiety of being apprehended and put on a flight back to Turkey. For the last two years, she has lived in the shadows, avoiding contact with authorities and moving from one rented room to another. The stress of living in fear has taken a toll on her health. But despite having a family doctor (a general practitioner), she fears the ramifications of seeking help:

> We have a GP (general practitioner), but we don't go because we are scared of the police. I mean I haven't gone to the GP for the past two years because I was worried the GP may inform the police or they will ask me to sign on again. I was worried that the GP and the Home Office work together. So we stay away from everywhere. Whenever I was ill I would just take painkillers and go to sleep. But two weeks ago I was really bad and I had to go to the hospital. We went in fear that

something was going to happen; they gave us tablets and painkillers to use for three days. We took them, and that was it.

The legacy of California's Proposition 187

In the 1990s, demographic shifts in the United States resulting from labor demand and US border policy that made it difficult for Mexican migrants to make return trips home resulted in sizeable increases in the number of undocumented migrants (see Massey, Durand, and Malone 2002). Much of the public discourse at the time suggested that the US–Mexico border was out of control and that undocumented migrants in the United States were draining public resources.

Demographic change driven by migration was perhaps most felt in California, where there were an estimated 1.3 million undocumented migrants, including a sizeable number of undocumented children. In 1994, Republican Assemblyman Dick Mountjoy introduced Proposition 187, a ballot measure to deny public services such as public education and health care to undocumented migrants. The "Save Our State" initiative, as it was called, would also require all state and local government employees to check the immigration status of those with whom they interacted and report suspected undocumented migrants to the Attorney General's Office. The idea was to make it difficult for undocumented migrants to persist in the state.

California Governor Pete Wilson used Proposition 187 as a central issue in his re-election campaign, evoking images of a tidal wave of migrants crashing down on the Golden State and draining public services. California residents ultimately passed the initiative by a wide margin: 59 percent to 41 percent. However, the law never went into effect. It was quickly challenged in a legal suit and was found unconstitutional by a federal district court. Nevertheless, Proposition 187 served as a prologue for similar actions decades later.

More than fifteen years later, in 2010, Arizona became the first state in the United States to enact far-reaching anti-immigrant legislation when it passed Senate Bill 1070. Its enforcement provision

– commonly known as "Show me your papers" – resulted in a legal justification underpinning racial profiling.³ Briefly put, the bill required state law enforcement officers to determine a person's immigration status during a stop or arrest when there was suspicion that the person in question is an undocumented migrant. In essence, the bill conflated Mexican nationality, immigrant status and illegality.

Following Arizona, several other states passed bills to levy sanctions against employers and empower local police to check for citizenship status.⁴ In 2011, lawmakers in Alabama went considerably further with House Bill 56, which sought to deny immigrants access to virtually every facet of regulated life, from water utilities to rental agreements to dog tags (Johnson 2011; Robinson and Preston 2012). Across the country, an even larger number of municipal and county ordinances attempted to cut off access to a wide array of common services (Walker and Leitner 2011).

Policies like these state measures are typically presented as targeting solely undocumented migrants. But if one considers the linkages migrants possess – to family members, neighbors, communities of faith, and employers, for example – it makes sense to suggest that, in reality, they have a blanket effect, directly or indirectly affecting whole communities. As the United Kingdom's hostile environment policy cast a wide net of enforcement, many communities experienced the collateral effects.

UK's hostile environment policy

Across the Atlantic in 2010, the UK Home Office initiated a set of administrative and legislative measures called the "hostile environment policy." Similar to the aforementioned state measures in the United States, the aim of the policy according to Home Secretary Teresa May was "to create, here in Britain, a really hostile environment for illegal immigrants." The policy included the removal of homeless EU citizens, requirements for landlords, healthcare professionals, charities, and banks to carry out identification checks.

The policy also implemented a more complicated application

Enforcement, Detention, and Deportation

process to obtain a "leave to remain," based on the principle of "deport first, appeal later" and additional efforts to encourage voluntary deportation. Under its directives, UK border officers were deployed to Chinese supermarkets, Indian restaurants, and kebab shops in East London and Brixton in search of undocumented migrants, and to wedding ceremonies of people with South Asian sounding names in search of immigrants committing marriage fraud (Jones et al. 2017).

In 2018, due in large part to the hostile environment policy, dozens of people were detained by the Home Office and denied legal rights in what was called the "Windrush scandal." All told, some 63 people were wrongly deported. In addition, an unknown number were wrongly detained, denied medical care or benefits, and lost their jobs or homes as a result. What's more, many long-term residents were denied entry to the United Kingdom and many more were threatened with deportation orders by the Home Office.

The vast majority of those affected by the policy had been born British subjects prior to 1973 and were Caribbean in origin as members of the "Windrush generation" (named after a ship that brought one of the first large groups of West Indian migrants to the United Kingdom in 1948). These individuals had exercised their right to "freedom of movement" within the borders of British territories.[5] Because their status enabled British subjects to legally reside in the country, many of them never applied for a British passport. However, the rules of residence have since changed, and many who came as British citizens are now treated as migrants. And, since the hostile environment policy, many have been put out of work, denied access to medical treatment, and even deprived of their pensions. In this sense, the hostile environment policy is pervasive and operates as a logic of governance that reshapes the rights of citizens and migrant alike and redefines the meaning of citizenship in the process (Sigona 2018).

These examples from the United States and the United Kingdom highlight an additional and important point: immigration enforcement is seldom color-blind. Immigration controls often act on stereotypical and racialized representations of illegality. As a result, certain demographic groups, due to their race or religion, are more visible and as a result more frequently targeted, while others often go unnoticed.

The deportation–detention nexus

According to official data from the US Department of Immigration and Customs Enforcement (ICE), nearly three million people were deported from the United States between 2008 and 2015, averaging almost 360,000 removals per year. Similarly, across the European Union more than 200,000 non-EU citizens were removed in 2017 (Eurostat 2018). It is worth noting that in both the United States and the European Union a larger number of migrants receive removal orders that, even though they are never acted upon, nonetheless contribute to creating an unstable and hostile environment for them and their families.

Over the last two decades, western governments have increasingly used denials of entry and forced removals as deterrents, while also assisting voluntary returns as instruments for governing unauthorized immigration. In addition, private immigration detention facilities have become big business and sophisticated and intrusive technologies have been employed by growing numbers of state and private border guards to identify, restrict, and remove thousands of people. In fact, detention and deportation have become "interlocking industries" of immigration enforcement (Mountz 2012: 523), as migrants facing removal orders increasingly interact with both systems.

"The formidable deportation machine"

Historically, deportation has served as an instrument of defining insiders and outsiders, while setting the parameters for inclusion and exclusion within countries' boundaries. Increasingly, however, mass detention and deportation have become normalized as a tool for migration governance (Bloch and Schuster 2005). This process has been aided by a compulsion towards the securitization of migration that has picked up pace since 2001 and the development of what former US Immigration and Naturalization Service Commissioner and her colleagues have called the "formidable deportation machine (Meissner et al. 2013; Rosenblum et al. 2014).

Coinciding with the rise in the size of settled undocumented populations in western countries has been the creation of a greater-reaching and arguably more punitive enforcement infrastructure. In 1996, the US–Mexico border became highly militarized, with the erection of a new fence, and renewed efforts to drive migration away from certain crossing points (Nevins 2002). That year, US President Bill Clinton signed a pair of bills which imposed stronger penalties for unauthorized entry and provided broader latitude to expel undocumented migrants by restructuring the detention and deportation process and through expedited deportation proceedings (De Genova 2002, 2005; Inda 2006; Macías-Rojas 2016). Taken together, the Antiterrorism and Effective Death Penalty Act (AEDPA) and the Illegal Immigration Reform and Immigrant Responsibility Act (IIRAIRA) expanded criminal grounds for deportation – including minor drug offenses, petty theft, and driving under the influence – and made deportation mandatory for all immigrants sentenced to a year or more.

In addition, the 1996 laws eliminated a practice known as "suspension of deportation," which had previously shielded immigrants who did not have a criminal history. In effect, this increased the number of noncitizens eligible to be deported while also decreasing their eligibility for relief from deportation. Due to the retroactive nature of the provision, immigrants convicted of crimes and those with past criminal convictions were subject to mandatory detention and deportation. It also left immigrants without the right to judicial review or appeal.

Understanding the relationship between mass incarceration and immigration enforcement provides some additional insight into the acceleration of detention and deportation efforts (Golash-Boza 2015a; Macías-Rojas 2016). The expansion of the aggravated felony category coincided with overcrowding issues in prisons during the US "war on drugs." By deporting undocumented migrants who were serving jail time, this new policy served to free up needed space in prisons (Macías-Rojas 2016).

Prior to 1996, deportations typically involved the voluntary returns of immigrants apprehended at or near the border. Removal proceedings from the country's interior were relatively rare and were subject to judicial review. Since 1996, this infrastructure has been strengthened by increased staffing levels for the US Customs and Border Protection (CPB) for border enforcement and the ICE for interior enforcement. In addition, increased integration between local law enforcement and ICE has expanded the capacity to identify and remove deportable immigrants.

Following AEDPA and IIRAIRA, in 1998 Congress phased out a provision of the Immigrant National Act (INA), known as Section 245(i), which permitted migrants to adjust their status within the United States (Gonzales 2016). Instead, migrants were required to return to their countries of origin. However, on leaving the United States, they would trigger mandatory bars to admission of three to ten years. These policies had the unintended consequences of de-incentivizing departures from the country and, as a result, greatly expanded the number of settled undocumented migrants living in the United States.

In addition, the federal response to the events of September 11, 2001 further expanded efforts to secure borders and increase enforcement. Guarding against the threat of terrorism, immigration enforcement expanded dramatically through a sixfold increase in raids, detentions, and deportations (Golash-Boza 2012). In effect, the adoption of stricter measures, under the auspices of national security, has drawn bolder lines around criminality and illegality. These measures have also aggravated demand for manufactured identity documents and driven undocumented labor further underground.

Criminal alien program and prevention through deterrence

Following a sweeping set of policies that criminalized undocumented migrants, the US Congress initiated another set of measures that would dramatically expand the reach and capacity of immigration enforcement. Taken together, the Criminal Alien Program (CAP), Secure Communities, the 287(g) program, and the National Fugitive Operations Program (NFOP) moved immigration enforcement into local communities by enlisting local law enforcement to aid with the detention and deportation of "criminal noncitizens" (Golash-Boza 2015b; Macías-Rojas 2016). In 2011, Congress allotted US$690 million for these four programs, a 30-fold increase from the US$23 million allotted in 2004 (Golash-Boza 2015b). Of the four programs, CAP has had the largest impact on deportations. From 2010 to 2012, the program generated half of the 400,000 deportations issued by the Department of Homeland Security (Golash-Boza 2015b). Importantly, CAP has served to criminalize daily life (Golash-Boza 2015b); Macías-Rojas 2016).

The Criminal Alien Program prompted the implementation of large-scale initiatives in several border-states, including Operation Gatekeeper in California, Operation Hold the Line in Texas, and Operation Safeguard in Arizona. These efforts dramatically militarized the US–Mexico border and impacted migrant safety (Dunn 1996, 2001; Nevins 2002; Rosas 2012). In addition to increasing the number of border patrol agents along the border, these efforts also initiated the use of surveillance technology – including helicopters, low-level light cameras, and radar equipment – to identify, track, and apprehend undocumented migrants along the migrant trail (Macías-Rojas 2016; Rosas 2012).

Not surprisingly, these measures had adverse effects. The fortification of certain areas of the border, largely urban, served to deter migration to areas more difficult to patrol and drove up human smuggling efforts (Andreas 2009; De León 2015; Rosas 2012). These practices, commonly known as Prevention through Deterrence, made border crossing much more dangerous and increased the risk of injury and death (De León 2015). For many migrants attempting to make the journey through these harsher environments, like the Sonora Desert, death has become an increasingly common outcome (De León 2015).

Border towns, like Nogales, Arizona, have experienced some of the most deleterious effects of measures such as CAP. Combined with efforts to beef up the US–Mexico border, enforcement measures in towns like Nogales span border crossings *and* everyday life. ICE agents and local law enforcement officers now share information with each other. And communities in these areas have seen a sizeable increase in the number of detention centers. As a result, a larger range of mundane everyday activities can land migrants in the immigration system.

As large-scale programs and policies have made their way into the interior and effectively extended the border into other areas of life, cities, towns, and neighborhoods have become central sites for the targeting of undocumented migrants who have settled in these areas. Increased cooperation between ICE and local law enforcement agencies, made possible through the implementation of 287(g) and Secure Communities, has transformed state and local law enforcement into immigration enforcement mechanisms, and has greatly facilitated the identification and apprehension of deportable immigrants and legal permanent residents with criminal histories.

The creation of 287(g) agreements, named for the section of the Immigration and Nationality Act that authorized such agreements, provided immigration-related training to local law enforcement and delegated authority to local law enforcement agencies to enforce immigration law within their jurisdiction (Chacón 2012; Leerkes, Bachmeier, and Leach 2013; Provine et al. 2016). These efforts were enhanced and enlarged by the adoption, across the United States, of the federal Secure Communities Program. Secure Communities allows law enforcement to forward fingerprints collected in the booking of those arrested to the Department of Homeland Security. According to the US Immigration and Customs Enforcement (ICE) webpage, Secure Communities has been fully implemented in all 3,181 jurisdictions in 50 states, the Washington DC area, and five US Territories since January 22, 2013 (ICE 2018b).

In doing so, these programs have transformed routine traffic stops and checkpoints into sites of deportability and criminalization. Concretely, this means that migrants have been arrested, detained, and deported following improper lane changes and broken tail lights, for example, causing devastating ripple effects on their families and communities. When one considers that the majority of US states do not permit undocumented migrants to acquire driver's licenses, their vulnerability is greatly enhanced (Coleman and Stuesse 2016).

In conjunction, these measures have served to increase the number of formal removals, increased deportations resulting from interior enforcement, and substantially increased the costs and sanctions associated with reentry into the United States. They have dramatically transformed immigration enforcement practices and also life in the United States for undocumented migrants and their families. The Obama administration chose to suspend Secure Communities on November 20, 2014, after public scrutiny for the program rose and local officials refused to cooperate with the program (ICE 2018b; Johnson 2014). However, it was reestablished by the Trump administration on January 25, 2017.

During the Obama administration, the use of prosecutorial discretion surfaced as an attempt to appease the backlash the administration received over their immigration enforcement agenda. Prosecutorial discretion uses the discretion authority of an officer or official in deciding which cases to pursue or not (Zatz and Rodriguez 2015). The use of prosecutorial discretion prioritized deportation cases with the aim of keeping migrant families together and aiding vulnerable migrant children (Zatz and Rodriguez 2015). Arguably, the results of using prosecutorial discretion have been mixed. Although efforts like prosecutorial discretion might have potential, it is important to recognize that it might take more than one policy to change or interrupt the production of illegality. Measures like prosecutorial discretion might help keep some families together and give individual migrants a chance to stay but they also do little to change overall immigration enforcement policies and the ripple effects they can have.

Since 1996, there have been several notable changes that have widened the deportation landscape: (1) the deportation process has been significantly streamlined; (2) the grounds for deportation have been greatly expanded; (3) interior enforcement activity has dramatically increased; (4) detention and deportation have occurred increasingly as the result of information gathered from local criminal justice enforcement (Gonzales and Raphael 2017). Taken together, these developments have increased the total number of removals from the country, increased deportations resulting from interior enforcement, and greatly increased the costs and potential sanctions associated with reentry. As such, they have fundamentally transformed US immigration enforcement and have dramatically altered daily life for undocumented migrants.

Over the last decade, hundreds of thousands of immigrants have been placed in removal proceedings after being stopped for minor traffic violations such as right turns on red lights, U-turns, and failing to use a turn signal when changing lanes or turning (Alonzo et al. 2011). In fact, between 1997 and 2012, the US government carried out more than 4.2 million deportations – twice the total number of all deportations from the United States prior to 1997 (Golash-Boza and Hondagneu-Sotelo 2013). In 2013 alone, the United States deported a record 438,421 immigrants (Gonzalez-Barrera and Manuel-Krogstad 2014). In fact, during the Obama presidency, more than 2.7 million immigrants were removed from the United States). These practices have had a strong chilling effect in communities across the United States.

Immigrant detention: A multibillion-dollar industry

Recent developments in immigration enforcement have seen a shift from a "catch and release" strategy to "catch and detain" (Douglas and Sáenz 2013), as the use of detention facilities has in recent decades proliferated on both sides of the Atlantic. This is evidenced by increased government funding for detention bed space and the proliferation of private detention centers (Macías-Rojas 2016).

In the United Kingdom, detention centers now number in the hundreds (Mountz et al. 2012; Schuster 2005, 2011). In the United States, immigration detention is the country's fastest-growing prison system. This is due in large part to the confluence of the growth of the private prison industry, the use of jails for immigration purposes, and the increasing convergence between immigration policy and the criminal justice system. Since the 1990s, the size and cost of immigration detention have skyrocketed. In fact, between 1994 and 2003, the number of those detained more than tripled in size, from 6,200 to 22,000 daily. What's more, the total number of people placed in detention annually increased from 85,000 people in 1995 to a peak of 477,523 in fiscal year 2012. While detention numbers had recently begun to decline, events at the US–Mexico border involving Central American migrants has dramatically elevated the number of detained migrant children. Population levels at federally contracted private shelters for migrant children increased fivefold between the summer of 2017 and roughly a year later, reaching a total of 12,800 in September 2018. Indeed, the separation of migrant children from their families in 2018 prompted protest around the globe. But the sizeable increases in child detention are not merely the result of a massive influx of children entering the country. Rather, this buildup also stems from the reduction in the number of children being released to live with family members or other sponsors.

For countries seeking greater levels of enforcement, the detention of immigrants provides several purposes as it functions as both an enclosure within a larger space, such as a prison or refugee camp, and an exclusion from the broader society (Bloch and Schuster 2005). Keeping migrants in detention facilities also ensures that migrants are present at their deportation hearings and that those contesting their immigration cases are warehoused while they move through the appeals process (Inda 2006). But for migrants, detention represents a loss of freedom, separation from family members, loss of work and family income, and a threat to safety and well-being. A recent report by ICE on immigrant detention found that in addition to being held in inappropriate environments, detainees had suffered physical and sexual abuse, lack of access to adequate medical care, and inadequate nutrition and exercise (Speri 2018).

Detainees include asylum seekers fleeing war or violence or political or religious persecution. They are unauthorized migrants apprehended at the border or within the interior of the receiving country. And they are those individuals sought out by government authorities because of their religion or nationality. In Europe, some countries detain asylum seekers upon arrival, while others detain them upon rejection of their asylum claim. While the United Kingdom, the Republic of Ireland, Denmark, and Greece have no established maximum time limit on stays, other countries' limits vary. For example, France's maximum time of detainment is 32 days, Italy's is 60 days, and Germany's is up to six months (Schuster 2005). But while France has the strictest limits among EU countries, conditions within its detention facilities are arguably the worst (Schuster 2005). Visitation in most of the detention facilities throughout the European Union is also commonly problematic due to noise levels in the facilities. Issues of exclusion and confinement are also present due to uneven access to legal representation and the detention of children (Bloch and Schuster 2005).

According to the US federal government, more than 60 percent of those held in immigration custody are held in privately run detention centers. In 2017, the two largest contracts with ICE were GEO Group and Corrections Corporation of America, receiving US$184 million and US$135 million respectively. The privatization of detention facilities has been linked to the emergence of what many have termed the "immigration industrial complex," a convergence of interests in the criminalization of undocumented migration and the compulsion to exert control over undocumented migrants through detention and deportation (Doty and Wheatley 2013; Golash-Boza 2009; Douglas and Sáenz 2013). This complex includes private military and security contractors that sometimes manage detention facilities and develop surveillance software and infrastructure (e.g., walls and razor wire), as well as a multitude of intergovernmental and nongovernmental agencies involved. Private corporations now play a major role in detention efforts by profiting from contracts with governmental agencies to run private detention centers (Douglas and Sáenz 2013). This industry directly profits from the construction of migrants as threats. As such, increased apprehensions, either at the border or within a state's territory, translate to increased profits (financial or otherwise) for the multitude of actors and agencies involved.

Practices of immigrant detention also reinforce constructions of citizenship through denial and affix categories of exclusion upon migrants (Mountz et al. 2012). Migration scholar Stephanie Silverman (2012) argues that immigration detention represents the "deprivation of a noncitizen's liberty for the purposes of an immigration-related goal, with 'noncitizen' signifying a 'person subject to immigration control'" (Silverman 2012: 1134). Research carried out on asylum seekers in the United Kingdom, France, Italy, and Germany suggests that detention has an adverse effect on mental well-being (Fekete 2005; Schuster 2005; Welch and Schuster 2005). This is manifested in suicide attempts and other forms of self-harm.

Detention also allows for the proliferation of ideologies casting migrants as "illegal criminal aliens" (Mountz 2012). Criminologists Michael Welch and Liza Schuster argue that the detention of asylum seekers not only reflects a "criminology of the other" but also disregards the mental and physical state of the detainee and alludes to the "cultural disposition that supports a greater reliance on prisons while ignoring the long-term detriment that mass incarceration imposes on society, communities and prisoners" (Welch and Schuster 2005: 332).

Of the many recorded abuses and violations in detention centers, sexual abuse represents one of the most severe. In the United States, all 50 states, the District of Columbia, and the federal government impose criminal liability on staff who have sexual contact with detainees. A recent investigation into such abuses within US detention facilities found that there were more than 1,400 filed allegations of sexual abuse between 2012 and 2018, with 237 allegations in 2017 alone. In March 2018, US Immigration and Customs Enforcement ended a general practice of releasing pregnant women facing deportation. Prior to the directive, pregnant migrants being detained were permitted to be freed on bond or supervised release. The new directive mandates that ICE keep those in detention.

While growing attention has been paid to the evolving practices of detention, these examples underscore the urgent need to better understand its gendered dimensions, particularly as many women remain in their roles as mothers within the exclusionary space of detention. And in light of recent deaths of two Guatemalan children in US detention facilities within a three-week time period, the confinement of children is particularly problematic (Jordan 2018). In countries that do not recognize birthright citizenship, where children who were born and raised in those countries face the troubling prospect of removal, families are at times held in "special exit centers," induced to embark on their voluntary departure (Fekete 2007).

Conclusion

This chapter has examined immigration enforcement practice and the impact it has on undocumented migrants, their families, and the communities where they reside. We began by outlining "soft enforcement" measures on both sides of the Atlantic. These practices are put in place by governments in order to create hostile environments for undocumented migrants, which are expected to act as deterrents for aspiring new migrants and to facilitate voluntary returns. We subsequently focused on the detention–deportation nexus and the massively profitable industry that has grown around it.

We looked then at how contemporary methods of enforcement dislocate immigration controls away from the geographical and juridical space of the border and towards the interior of the nation-state, in cities, towns, neighborhoods, workplaces, hospitals, schools, and public spaces. The stories in this chapter illustrate how these measures sow fear and anxiety in families and across communities and transform daily activities like taking the bus, dropping children off at school, and walking to traumatic events like arrest, detention, and deportation. In this sense, these measures deplete workforces, disrupt community harmony, and separate families. One such example in the chapter is Arizona's "Show me your papers" directive, which requires state law enforcement officers to determine a person's immigration status during a stop, and to arrest those under suspicion of being undocumented.

Similarly, under the United Kingdom's hostile environment policy, border officers were deployed to Chinese supermarkets, Indian restaurants, and kebab shops, mostly in East London and Brixton, in search of undocumented migrants working there. The detrimental impact of such hostile and discriminatory environments in the United Kingdom was in full view in 2018, when dozens of people were detained by the Home Office and denied legal rights in what was called the "Windrush scandal."

Through these examples from both sides of the Atlantic we noted that immigration enforcement is seldom color-blind. To the contrary, immigration controls often act on stereotypical and racialized representations of illegality. As a result, certain demographic groups, due to their race or religion, are more visible and as a result more frequently targeted, while others often go unnoticed.

Finally, the chapter highlighted how over the last two decades, alongside such soft measures that entrench the border in everyday spaces and induce anxiety into undocumented migrants and entire communities, western governments have increasingly deployed detention and deportation for governing unauthorized migration. In fact, recent developments in immigration enforcement have seen a shift from a "catch and release" strategy to "catch and detain." Consequently, private immigration detention facilities have become big business, and sophisticated and intrusive technologies have been employed by growing numbers of state and private border guards to identify, detain, and remove thousands of people. In fact, detention and deportation have become interlocking industries of immigration enforcement, as migrants facing deportation are increasingly detained for long periods of time. In conjunction, these two sets of measures have served to substantially increase the number of apprehensions and subsequent detentions and deportations.

5
Undocumented Status and Social Mobility

One of the biggest reasons I learned I slipped through the cracks at Goldman Sachs and why I never got caught or deported is because we have this very narrow-minded view of who an undocumented migrant is and what they do, and I didn't fit that stereotype, so when people were looking at my papers they were never questioning are these papers real or not?

Julissa Arce, 2016, *My (Underground) American Dream*

We're not always who you think we are. Some pick your strawberries or care for your children. Some are in high school or college. And some, it turns out, write news articles you might read. I grew up here. This is my home. Yet even though I think of myself as an American and consider America my country, my country doesn't think of me as one of its own.

Jose Antonio Vargas, 2011, *My Life as an Undocumented Immigrant*

We are the people you do not see. We are the ones who drive your cars, clean your rooms.

Dirty Pretty Things

When she was 11 years old, Julissa Arce left Mexico to join her parents in San Antonio, Texas. She entered the United States on a tourist visa, but when the visa expired when she was fourteen, she became undocumented. Julissa's parents had limited means to support her education. But they insisted she attend private school, knowing she would have greater opportunities. Both parents worked tirelessly to provide for the family, selling silver at US trade shows, then selling funnel cakes and snow cones as street vendors.

Good grades in high school and favorable policy in Texas at the right time facilitated Julissa's entry into the University of Texas. In 2001, the year of Julissa's high school graduation, Texas passed legislation allowing undocumented migrants like her to pay tuition at in-state rates and to receive state financial aid. She worked her way through college, taking over her parents' food cart. Eventually, she wanted better job opportunities – employment that matched her education and talents. So she purchased a fake green card and social security number.

Shortly after graduating with her bachelor's degree, Julissa landed an internship at Goldman Sachs. Six years later, she was earning more than US$340,000 annually as a vice president within the company. However, Julissa's handsome salary came at a price. She was working 80-hour weeks, and she lived in constant fear of her immigration status being discovered.

By 2009, Julissa had grown tired of the grueling pace and of hiding. She eventually left Goldman Sachs. And, as fate would have it, she was able to obtain permanent residency through marriage.

Two years later, another successful undocumented migrant very publicly disclosed his status. In an article written for the *New York Times Magazine*, the Pulitzer prizewinning journalist Jose Antonio Vargas described his life in the United States, how he discovered as a teenager that he was undocumented, and how he continued working towards his dreams while still constantly worrying about the implications of being caught.

Vargas's narrative received considerable attention, and for several years it did not have a visible impact on his ability to live, work, and travel in the United States. This was particularly striking during a decade of enforcement efforts that removed hundreds of thousands of immigrants annually. That is, not until July 2014 when Vargas was detained near the US–Mexico border in McAllen. Vargas was visiting a shelter for unaccompanied Central American children to share his story about his own journey to the United States when he realized that he might have difficulty in this border region getting back on a plane to return home with his passport from the Philippines. Because of the active intervention of his politically connected colleagues, Vargas was released, but not before he had been handcuffed and detained.

Why begin this chapter with the stories of a talented financial analyst and a celebrated journalist? Our intention here is twofold. First, we wish to continue to broaden the public imaginary regarding who is an undocumented migrant. Second, we want to illustrate the diverse pathways of social mobility among undocumented migrants. Indeed, undocumented status dramatically circumscribes opportunity, and many undocumented migrants live and work at the margins.

But undocumented migrants are not a monolith of individuals sharing the same national, racial, class, and educational backgrounds. Like any other population, undocumented migrants are diverse in both the skills they may bring from their home countries and the opportunities and barriers they face as undocumented residents. Age, gender, race, immigration history, and social network, as well as the opportunity structure of the country of residence, all contribute to determine the contours of an individual's undocumented life. While many migrate with minimal levels of education, others possess advanced degrees and have access to certain resources. In addition, receiving contexts, at the national as well as the local level, vary greatly. Opportunities and barriers to social mobility also vary accordingly. For both Julissa and Jose, the specter of deportation did not foreclose educational and employment opportunities. But neither did their educational and social privilege fully protect them from state rules and regulations. For young undocumented migrants, access to non-compulsory education and a formal labor market is often precluded and, as a result, chances of social mobility are dependent on achieving some form of status regularization.

How does undocumented status shape social mobility? And how is the social mobility of undocumented migrants mediated by other demographic characteristics, experiences, and contexts? This chapter explores the contexts shaping different life trajectories and, in doing so, also assesses the limitations of inclusionary contexts.

Pre-migration factors determining starting points

Much of the research on immigrants' social mobility has focused on either their cultural values or structural barriers in receiving countries. For example, do the children of certain immigrant groups succeed in school because their families place more value on school achievement, or do inequalities in opportunities and resources structure disparate outcomes? One often overlooked factor explaining disparities in mobility is the differences in socioeconomic conditions prior to migration (Feliciano 2005, 2006). Coming from a higher-class position in one's home country, for example, may give some immigrants advantages that are not usually captured by traditional measures of status in the receiving-country context. These advantages might also provide a significant boost to their children in the receiving country.

While some immigrants arrive with visas, advanced degrees, and well-paying professional jobs, others come clandestinely, with minimal levels of education, and find employment in low-wage labor markets. This diversity of immigrant origins contributes to divergent "starting points" from which they begin their lives in new countries (Lee and Zhou 2015).

As Cynthia Feliciano points out, immigrants are not random samples of their home countries' populations. Instead, they originate from particular segments of their sending societies (Feliciano 2005, 2006). Indeed, only some people want to migrate or have the resources to do so. In the case of forced migration, some migrants have more resources that influence their migration journeys and their settlement in receiving countries. Migration is also shaped by political and economic contexts in sending countries, historical relationships between sending and receiving countries, and the receiving country's immigration policy (Massey 1999; Menjívar 2000; Rumbaut 1997).

Indeed, what scholars call "immigrant selectivity" affects many facets of life, including immigrants' earnings (Borjas 1987; Carliner 1980; Chiswick 1991) and health outcomes (Landale, Oropesa, and Gorman 2000). As such, while individual migrants' backgrounds importantly shape their experiences in host countries, group-level characteristics can play an even greater role (Feliciano 2006; Portes and Rumbaut 2001, 2006). For example, one's immigrant community may be positively selected with respect to higher education level and thus may have a positive effect in shaping high expectations for their children around education (Portes and Zhou 1993; Zhou and Bankston 1998). Immigrant communities may also structure opportunities for co-ethnic migrants in the form of access to employment through local businesses and ethnic niches, community institutions, and advocacy.

Similar to legally residing immigrants, undocumented migrants migrate with a diverse set of educational and class backgrounds. To be sure, many undocumented migrants who undertake clandestine journeys and make unauthorized crossings have lower levels of education and possess fewer resources. Many of them live in communities composed of low-wage migrants and characterized by abject poverty and do not have the resources to help each other (Menjívar 2000). But many undocumented migrants also overstay visas and, as such, possess the requisite connections and resources to secure visas. Although they may be undocumented, they may have certain advantages – in the form of language skills, educational attainment, social networks, or material resources – that may allow them to access better employment options and resources unavailable to their less educated counterparts.

Local contexts structure opportunities

Sociologist Alejandro Portes and his colleagues argue that experiences of integration and exclusion are structured by the interplay between individual characteristics, on the one hand, and the structural and historical contexts shaping opportunity, on the other. In essence, while human-level variables, such as education, skills, and job experience, position individuals in different capacities to respond to opportunities within the labor market, one's "structural embeddedness" can constrain or boost individual action (Portes and Rumbaut 2001: 312). Accordingly, the context of reception – made up of the policies of the receiving government, the host society's reception of newcomers, the conditions of the labor market, and the characteristics of one's own ethnic community – determines the extent to which individual skills can be put into play (Portes and Rumbaut 2001: 313–14). As such, the combination of enabling and restrictive contexts determines one's distinct mode of incorporation. Since modes of incorporation can facilitate, alter, or even prevent the deployment of individual skills, they yield constraints and opportunities (Waldinger and Catron 2016).

The modes-of-incorporation framework is useful for examining the different ways in which individuals experience undocumented status. On the one side, national laws and enforcement practices substantially shape the hard boundaries of everyday life. However, local policies – whether exclusionary or integrative – can either reinforce these boundaries or soften them. In addition, community institutions, civil society, and foundations, where present, can buffer exclusionary policies and provide opportunities for immigrants to more fully engage with their communities. Belonging to a longer-standing immigrant community with a host of immigrant-serving organizations can create a strong sense of belonging and grant immigrants access to resources, opportunities, and other webs of support. By contrast, living in a new destination of settlement can present daily challenges, including hostile attitudes towards newcomers, a lack of resources and support, and restrictive local policies which can heighten the effects of illegality and deportability.

Indeed, recent scholarship has uncovered layers of stratification within undocumented populations, differences structured by race, place of residence, and education. However, some scholars maintain that illegality is a "master status" (Gonzales 2011, 2016). Sociologists Roberto G. Gonzales and Edelina Burciaga suggest that these two ideas are not in tension (Gonzales and Burciaga 2018). Drawing on the original master-status concept, initially proposed by Everett Hughes (1945) and further delineated by Howard Becker (1963), they point out that certain traits can be master statuses while also exhibiting layers of stratification. That is, most statuses have a primary trait, which distinguishes those who belong in the group from those who do not. But other auxiliary traits are often associated with the master status.

Undocumented migrants often face a uniform set of exclusions: in most countries, they cannot work legally, they are ineligible to vote, and are ineligible for most government-funded public insurance programs. They can also be apprehended, detained, or deported at any time. But, as we have previously argued in this book, undocumented migrants are not solely defined by their immigration status. They possess other traits that make up who they are and which either impede or facilitate opportunities in their daily lives. These traits might be otherwise dominant, or "master statuses" themselves, but become subordinate to their undocumented status. They might also be auxiliary traits – like being poor, low-wage workers, minimally educated – that are typically associated with undocumented migrants. So when undocumented migrants lack some of the auxiliary traits associated with undocumented status – they might have college degrees rather than be minimally educated – they may, then, possess some degree of status inconsistency. That is, while undocumented status operates as a master status, other traits also shape their lives.

For undocumented migrants, several factors – including pre-migration capital, local policies, the institutional makeup of a particular geography, and social connections – can play an important role in shaping opportunities and experiences.

Local policies structure opportunity and vulnerability

For many undocumented migrants, pre-migration advantages or disadvantages shape their starting points in receiving countries. However, their experiences of settlement are also largely formed by local contours, particularly local policies.

As immigration policy has become increasingly localized, the places immigrants live powerfully shape opportunity. Due to a constellation of local policies, the experience of undocumented migrants in Los Angeles can differ from those in Charlotte, North Carolina, let alone in Athens, Rome, or Birmingham, England. In the United States, local policies towards immigrants differ not just by state but also by county and municipality, while in the case of Europe they are often constrained by EU legislation.

As we discussed in chapter 3, the devolution of immigration policy in the United States to states, counties, and municipalities has resulted in an "uneven geography" of policies and practices (Coleman 2012). The lack of comprehensive federal immigration reform in the United States has put enormous pressure on states and localities to craft their own policies to address local immigration issues. The response by local governments has been varied and ever changing. Some cities have responded by creating inclusive policies that not only integrate undocumented migrants into their communities but also allow them access to resources critical to their social and economic well-being. Others have taken on more exclusionary stances further circumscribing the lives of undocumented migrants and dramatically reducing opportunities for social mobility.

This variance in immigration policy and enforcement at the local level has produced divergent experiences (Coleman 2012). For example, an undocumented migrant in Los Angeles may have access to a driver's license, in-state tuition and financial aid for public colleges and universities, and to a professional license to practice any of forty enumerated professions, yet be negatively affected by a high cost of living that soars above the national average.[1] A similarly situated undocumented migrant in rural Georgia may be negatively impacted by exclusions from medical care and post-secondary education, and increased integration between police activities and federal Immigration and Customs Enforcement (ICE) agents, but benefit from a strong network of community organizations.

Local immigration policies – whether integrative or exclusionary – provide opportunities for as well as impediments to social mobility. For example, legislation aimed at providing undocumented migrants access to driver's licenses opens up possibilities for employment and broader community participation. With the ability to lawfully drive, undocumented migrants can seek a broader range of employment opportunities across a wider geography not limited to options near public transportation routes or jobs closer to their homes. Having that confidence can also lend some stability to their work life, allowing them to stay in their place of employment and potentially achieve some mobility in that job over time. In addition, driving with valid licenses can also boost confidence to seek out opportunities without fear of being caught and detained. And it can provide the children of undocumented migrant parents a more expansive terrain from which to seek out educational enrichment opportunities. Similarly, policies that allow undocumented migrants to access post-secondary education give them a chance to invest in their human capital.

Conversely, restrictive immigration policies can substantially limit options for social mobility. Local policies to formalize cooperation between local law enforcement and immigration officers, combined with exclusions from driver's licenses can dramatically curb transportation options, as they have the potential to reduce unlicensed driving by undocumented migrants who fear that traffic stops could result in their deportation. This could, in turn, shrink their geographic options for work, schooling, and supplementary educational and employment opportunities. What's more, policies that exclude undocumented migrants from post-secondary education and job training opportunities put a ceiling on human capital development and social mobility.

The community–institutional context

The incorporation of undocumented migrants is a process bounded by place and is largely determined by access to social support services that can help alleviate the lack of legal status. However, many undocumented migrants lack access to the kinds of supports necessary for their social and material well-being. Additionally, many social services fall to localities to administer, making it more difficult for government-sponsored incorporation services to be readily available and placing the impetus on non-profit organizations, schools, or other social service organizations to manage the incorporation of newcomers. As noted earlier, localized incorporation efforts are not uniform and depend significantly on the various sociopolitical contexts of reception. Finally, as many scholars have illustrated (Chacón 2014; Chauvin and Garcés-Mascareñas 2014; Coutin 2013), these fragmented modes of incorporation reflect the ways in which illegality is produced and reproduced within frameworks of inclusion and deservingness.

Recent studies exploring the community contexts of immigrant destinations have expanded our understanding of the local context and its importance in the incorporation process for newcomers (Bickham and Nelson 2016; Dreby and Schmalzbauer 2013; Parrado and Flippen 2016; Lehman 2016). Of particular importance to migrants and their families are infrastructures of support: organizations that provide services in their languages, institutions that serve undocumented migrants, and accessible options for educational and social assistance. In addition, access to public transportation, affordable housing, and a welcoming community can facilitate integration. The constellation of positive or negative contexts within the local community can either undercut migrants' efforts to eke out a manageable life or provide them with a significant boost over certain barriers.

Many locales with long-standing migrant settlement possess the infrastructures necessary to provide more cohesive incorporation trajectories to new immigrants (Portes and Stepick 1993). These locales tend to be more urban and with high concentrations of diverse populations. As long-standing ports of entry for immigrant communities, they've seen waves of immigrants establish themselves, create institutions that serve their communities (e.g., banks, churches, businesses), and set up social service and educational organizations to meet the various needs of their residents. Over the years, these "enclaves" have absorbed continuous waves of immigrants and have provided them social assistance, employment opportunities, and community support (Portes and Zhou 1993). In addition, these communities tend to have more and better-established organizations and institutions that are well equipped to serve the diverse needs of newcomers.

In contrast, newer immigrant destinations in the United States tend to be in rural areas in the Midwest and South (Massey 2008; Zuñíga and Hernández-Léon 2005). What's more, immigrants in America's largest metropolitan areas are now more likely to live in the suburbs than in central cities (Allard 2009; García Bedolla 2005; Hall and Lee 2010; Holliday and Dwyer 2009). Research suggests that the availability of low-skill work and lower costs of living provide beneficial opportunities for migrants to weaken the economic barrier that separates them from the native population (Marrow 2011). In addition, immigrants' perception of these communities as safe, less violent, and slow-paced encourages permanent family settlement (Dalla, Ellis, and Cramer 2005). However, migrants in new rural and suburban destinations also encounter a negative context of reception that includes widespread poverty, limited opportunities for stable employment, a decrepit social service infrastructure, and a lack of public transportation options (Allard 2009). Although overall poverty rates tend to be high throughout rural America, Mexican migrants experience acute poverty at higher levels than any other ethnic or racial groups and are more likely to earn less than the median US income (Kandel et al. 2011; Lichter and Johnson 2009).

As the last two decades have seen growing numbers of migrants bypassing urban areas, the geography of poverty has also shifted. Low-income individuals in American metropolitan areas are now concentrated in suburbs rather than in urban areas. The confluence of these two shifts marks an unprecedented period: at no other time in history have American suburbs been home to a greater share of low-income and foreign-born residents relative to the urban core. Low-income suburban areas face challenges to economic and social incorporation that are often distinct from the experience of their migrant counterparts in urban immigrant neighborhoods. While they may have better access to low-skilled jobs, their low wages and the transient nature of agricultural and factory jobs limit opportunities for upward mobility. In addition, suburban communities do not typically feature a diverse set of ethnic organizations to help low-income newcomers adapt to their receiving context. This places an additional burden on local organizations such as schools, churches, social service organizations, and hospitals to meet the needs of the growing migrant populations.

There is a small but growing number of ethnographic studies focusing on the adaptation of immigrants in rural and suburban communities (Gonzales and Ruiz 2014; Roth 2015; Stein et al. 2016). These studies highlight the difficulties that organizations have in responding to the needs of migrants and their families (Allard 2009; Roth, Gonzales, and Lesniewski 2015). While these communities may host social service organizations, existing organizations can face numerous and striking challenges in providing support for low-income migrants, including finding bilingual staff and securing the funding to pay them. The ability to offer services in migrants' native languages is critical to immigrant-serving organizations and schools because having bilingual staff can help lower linguistic and cultural barriers.

In addition, undocumented migrants in particular communities experience racial segregation and racially charged attitudes. Such negative contexts of reception by the host society can hinder access to the kinds of resources needed to facilitate their incorporation and social mobility (Hall and Stringfield 2014). Undocumented migrants in newer destinations, such as Nashville, Tennessee and Charlotte, North Carolina, have experienced difficulty integrating into the local fabric of these locales (Furuseth, Smith, and McDaniel 2015). This difficulty is, in part, due to restrictive state policies that are meant to preclude undocumented migrants from incorporation and mobility opportunities. Despite the challenges they might face, undocumented migrants do find ways to build their own spaces of belonging and networks of support through entrepreneurial ventures, such as establishing ethnic food trucks, supermarkets, and other local businesses, as well as through collective action efforts (Furuseth et al. 2015; Okamoto and Ebert 2010). Moreover, through these ventures immigrants are able to contribute to efforts to revitalize local economies (Wortham, Mortimer, and Allard 2009).

Indeed, longer-standing migrant gateway cities like New York and Los Angeles are also characterized by racial and economic inequality. However, many of these larger urban centers possess the education and social service infrastructures critical in the incorporation process (Okamoto and Ebert 2010; Pastor and Mollenkopf 2012).

As to the effectiveness of institutional infrastructures at the nation-state level, an excellent case in point is the educational system of Greece. While Greece has been an immigrant-receiving country for generations, a more recent wave of migrants has settled there since the 2015 refugee crisis. When migrant children began arriving in Greece, they were obliged to attend school. But they did not speak the language and schools lacked staff that spoke the children's native languages. As a result, the migrant children were often neglected by teachers and bullied by their Greek peers. The government solved its dilemma resourcefully: it began employing asylum seekers and refugees who could communicate with the migrant children as well as others in their new schools. These individuals have served not only as interpreters but also as cultural mediators for government agencies and social services. In its medium-term-oriented planning, the Greek government established reception classes within the public education system and on school premises in order to prepare these newly arrived children for their eventual incorporation into the education system. However, these classes were held outside normal school hours so as to avoid disrupting the educational process of the students. This in turn created a ghetto-like situation in many Greek public schools, which further marginalized migrant children but also alienated local native communities.

The rising importance of place illustrates how pivotal the role of local organizations, schools, and other service infrastructures are for the incorporation of migrants. As the example of the Greek school system shows, these services can work to cushion the effects of illegality while allowing undocumented migrants not just to partake in community life but also to become part of its fabric.

Schools

Education scholars have long stressed the importance of schools as instrumental in the incorporation process for immigrants, undocumented or otherwise. Yet schools are often sites of stratification and inequality (Bowles and Gintis 1976; Giroux 1983).

In 1982, the US Supreme Court ruled in *Plyler v. Doe* that school districts could not deny access to the children of undocumented migrants based on their immigration status (Olivas 2012). Similarly, in Europe, all school-aged children, regardless of their residential status, have access to a free public education system. As a result, undocumented children are afforded the opportunity to be legally integrated into a powerfully defining institution.

Generally speaking, schools have long been viewed as mechanisms through which students gain much-needed tools to become productive members of society. Moreover, schools are instrumental in helping students become familiar with the social, cultural, and political norms of the nation-state (Abrego 2006; Gonzales, Heredia, and Negrón-Gonzales 2015; Suárez-Orozco, Suárez-Orozco, and Todorova 2008). As such, schools can serve as effective sites of inclusion by providing undocumented migrant students and their families access to community networks and other social services. In addition, the socialization process that students undergo in schools affords them the opportunity to expand their social circles and build networks through their interactions with school counsellors, administrators, fellow students, and parents.

In the United Kingdom, the principle of the best interests of the child dictates how various educators approach undocumented, or irregular, immigrant pupils.[2] A teacher interviewed by Sigona and Hughes (2012) in London provides some insight into how educators view the integration of migrant children in their schools. This attitude reflects a more general view in the sector, that is, that the lack of legal residence status should be of secondary importance and not lead to the exclusion of children from the education system. "We're always prioritizing the child. If you just want to do everything by the book – 'Okay are you legal here?' – you are not really helping the child. And that is the priority for us, that he or she feels safe and that he or she doesn't feel threatened by being sent home at any time." Despite pressure put on schools by UK officials to check students' identification cards and travel documents, teachers find it difficult to keep abreast of the changes and complexity of the immigration system and the specificities of individual cases. However, they can see the impact of these changes on the children firsthand. In particular, teachers in the United Kingdom have described seeing signs of distress and erratic behavior in their older students (Sigona and Hughes 2012).

Many schools in diverse British cities have adopted a general view whereby they see their role as education providers and not as immigration officers. This means that they accept papers that are given to them and try to be as flexible as possible with regard to the type of document that is presented without getting themselves into trouble. As one school administrator explained:

> I wouldn't know what a forged Italian passport, or a Portuguese, or a forged Spanish document would look like, it's not my job, you know, all I do is to say "What is your nationality?" They then have to show that documentation to the enrolment officer and they photocopy it as evidence ... so I wouldn't know what a forged passport would look like.[3]

Additionally, students with undocumented status are able to gain a more cohesive sense of belonging through their school community. Relationships with teachers and other school staff can provide them with the kind of support needed to feel stable and comfortable within their schools. And connections with peers can more readily integrate them into their classrooms and communities. Taken together, school settings, and the relationships forged in them, can help students to be embedded in school activities and to feel part of a larger community.

What's more, undocumented students also serve as bridge and broker to different communities and resources (Stephen Lee 2015). For example, undocumented students often bridge the gap between the native community and their parents, other family members, and their community more generally. Often, undocumented students serve as translators, mediators, or in other support roles to help their families "partake in US life" (Lee and Zhou 2015: 1407). Consequently, schools play a pivotal role not only in the incorporation process of undocumented students but also in the lives of their families. Students gain the knowledge to empower them to become civically engaged, despite their lack of formal legal status.

Nevertheless, many migrant children live in unsafe neighborhoods, experience multiple relocations, and reside in cramped living quarters (Gonzales 2016). In addition to the immediate effects that these unfavorable conditions might have on young people, they also limit their ability to do their best in school. The school setting, while integrative to a certain extent, can also function as a sorting mechanism that becomes increasingly stratified once students reach the upper-grade levels and whereby students, judged by their abilities and behaviors, earn labels like "good student" and "college material" or "lazy" and "troublemaker." Designations within the school's curricular hierarchy can have short- and long-term effects. In the short term, they can powerfully shape students' ability to access resources critical to their school success. In the long term, they can have profound implications for opportunities beyond school and for students' beliefs about their own abilities (Gonzales 2010a). For migrant youth of economically struggling families, negative school experiences can motivate educational exits, particularly for those seeking to enter the workforce to contribute to the household.

Access to the labor market and laboring opportunities

As described earlier in this book, global processes produce a demand for flexible, cheap, and deregulated migrant labor in western liberal democracies. However, the legal systems of nation-states and the legal categories under which immigrants are designated within a particular country's given immigration system contribute to produce the type of migrant that satisfies this demand and shapes the ways in which migrants are incorporated into local economies and participate in institutions and communities. Nevertheless, this is not a one-way process. Migrants themselves play an important part in their own incorporation trajectories not only through their struggles and claims but also through everyday living. Indeed, while certain designations bestow upon migrants the right to work, vote, receive benefits, and travel outside of the country of residence, and others frame their legal exclusion and enforced vulnerability, there is much more complexity involved.

Restive immigration policies in western nations have created a large and vulnerable workforce with few protections. As anthropologist Sarah B. Horton (2016) underscores, migrants experience poor working conditions in dangerous, even at times deadly, environments (see also Bloch and McKay 2016; Goldring and Landolt 2012; Holmes 2013). Much of this work is highly stratified, with jobs that require more autonomy and offer better pay, such as those in packing and processing warehouses, given to those with legal status, while the difficult, labor-intensive jobs are where the undocumented workers are typically concentrated. This hierarchy exacerbates the exploitation of the undocumented worker through what Elizabeth Fusell calls "deportation threat dynamic" (Fussell 2011). The deportation threat dynamic describes the ways in which undocumented migrants who are without work authorization are particularly vulnerable to wage theft and exploitation from employers who use workers' vulnerability against them, knowing they are less likely to report crimes against them.

The food industry has also impacted migration patterns and immigrant networks. In some cases, the need for cheap labor has transformed specific regions that were not historically migrant hubs. The United States South provides a useful example of the ways in which labor demands attract newcomers and transform entire communities. In her book *Scratching Out a Living*, anthropologist Angela Stuesse outlines how the growing demand for chicken in the United States has transformed the workforce in chicken-processing plants and changed the demographic landscape of communities in Mississippi (Stuesse 2016). She notes, "as our consumption of America's favorite white meat escalated, the poultry industry harnessed globalization's technologies and neoliberalism's labor control strategies and began recruiting immigrant labor at unprecedented rates" (2016: 10). The growing demand for immigrant labor has exposed the insidious relationship between demand for cheap labor and worker vulnerability (De Genova 2002). Thus the transformation of locales by these industries is undergirded by the exploitation and hyper-vulnerability of undocumented workers. These conditions also make it difficult for advocacy organizations to incorporate this new workforce into unions (Stuesse 2016).

Despite the challenges of unionizing undocumented workers, union organizers have found ways to incorporate undocumented workers into their ranks. This incorporation has come at a time when unions are on the decline and their survival depends, in part, on placing more attention on the sectors that are dominated primarily by undocumented workers (Milkman 2011). Unionizing efforts in the janitorial, hospitality, and retail sectors have helped bridge issues of labor rights and immigrant rights, further emphasizing the intersecting needs of undocumented migrants as both workers and people seeking pathways to citizenship (2011). While not all undocumented migrants join unions, other attempts have been made to help build awareness among undocumented workers. One of these is the establishment of worker centers which focus on providing aid for workers seeking employment while also raising awareness among the workers about their labor rights and the like (Milkman 2011; Purser 2009; Valenzuela et al. 2006). Unions and worker centers provide varying levels of aid, guidance, and stability to undocumented workers who often find themselves in less than ideal conditions due to their precarious status.

Connections

The concept of social capital – how individuals and groups invest in social relationships and share resources – is relevant to our understanding of the mechanisms that promote or impede social mobility among undocumented migrants. As the work of sociologist James Coleman underscored, social capital exists in the structure of relationships between and among people (Coleman 1988). The key characteristic of social capital is its convertibility. That is, it can be translated into other social and economic benefits. People can access social capital through membership in interpersonal networks and social institutions and pursue their goals by converting it into other forms of capital to improve or maintain their position in society. According to Pierre Bourdieu and Loic Wacquant, social capital is "the sum of the resources, actual or virtual, that accrue to an individual or a group by virtue of possessing a durable network of more or less institutionalized relationships of mutual acquaintance and recognition" (Bourdieu and Wacquant 1992: 11).

Social capital is an important concept in immigration research as it provides a useful framework for scholars to examine ethnic resources and immigrants' ability to tap into their relationships and networks in order to gain information and support. Scholars credit social capital with aiding migrants in finding housing, obtaining jobs, and seeking support. In their book, *Growing Up American*, Min Zhou and Carl Bankston III demonstrate how immigrant social capital engenders more positive outcomes for children. They explain the academic success of low-income Vietnamese youths by pointing to the ethnic community and its simultaneous promotion of success and enforcement of social controls. For the Vietnamese children in their study, social capital from the family and the ethnic community helps them to overcome a lack of human and economic capital.

As undocumented migrants begin to settle and create community for themselves in their new places of residence, they also begin to develop networks of support. In contrast to Zhou and Bankston, and other studies that identify immigrant social networks as viable sources of assistance, some scholars argue that networks can be reconfigured and fragmented, problematizing their durability as viable sources of assistance. Rather than examining networks' adaptive function as coping mechanisms to the receiving context, they focus on the effects of broader structures on social networks themselves. Cecilia Menjívar (2000) makes the case that poverty renders it difficult for immigrants to accumulate enough resources to help each other. She points out that various contexts structure opportunity for immigrants, thus determining the viability of immigrant social networks. Social relationships, Menjivar argues, do not exist in isolation from the structures in which the immigrants live. For some immigrants, these contexts are favorable and endow their ethic community with resources that benefit its members. But for other immigrants negative contexts (at the level of government, society, and the ethnic community) diminish the value of social networks. For the Salvadorans in Menjivar's study, poverty makes it difficult for them to accumulate enough resources to help each other. Indeed, networks develop out of need and in reaction to poverty, but those very structural conditions often serve to undermine networks.

Fear of detection and deportation makes undocumented migrants particularly cautious when it comes to friendships and relationships. While aware of the advantage that may derive from being widely connected, many feel unable to disclose their status to fellow workers and schoolmates for fear of exposing themselves and their families to moral judgment and, more concretely, to avoid the risk of being reported to authorities (Sigona 2012).

A sense of isolation is evident among the young undocumented migrants interviewed by Bloch, Sigona, and Zetter, particularly among women and newcomers from countries with less-developed community networks. Diana, a 28-year-old Brazilian who recently arrived in London, is one of them. She explains why making friends is not so straightforward if you are undocumented and how she has learned to cope with isolation:

> If you meet people, you can't tell them much, you don't know if you can trust them or not. Sometimes, the few people, the few times I talked about it, I talked to people who don't have documents either, I joked, "If you do something to me, I'll take you down with me. I know where you can be found" (laughs) [. . .] I've created a skin to protect myself because, like, it's only me, if anything happens I have to deal with by myself. Nobody is going to help me, so I kind of grew this skin, closed myself down not to be affected, to avoid problems as much as possible. (Bloch, Sigona, and Zetter 2014)

More recently, immigrant communities have leveraged social networks to build awareness of their struggles and have used them to develop advocacy campaigns. In doing so, they have elevated their platform to make certain rights claims that help them not only to be seen as active members of their communities but also to create bonds of solidarity with other immigrant communities and citizens (Ansley 2010; Nicholls 2016). This can be seen in the various immigrant social movements across recent history. For example, the Sanctuary movement of the 1980s brought together recently arrived Central American refugees, religious leaders, community activists, and various other community members to rally support against the deportation of Central American migrants fleeing wars and conflict while advocating for their recognition as refugees (Coutin 1993, 2000; García 2006). A more recent movement is the DREAMer movement that has rallied support for the DREAM Act with the hopes of achieving a path to legalization for the 1.5-generation of undocumented migrant youth who came to the United States as children with their families (Gonzales 2008; Nicholls 2013; Seif 2004).

As we will discuss in greater detail in chapter 7, the mobilization of migrant social networks can lead to immigrant collective action. In fact, it is one of the main strategies that allow undocumented migrants to forge a place for themselves while making claims for rights that can help them push the boundaries of illegality through resistance (Okamoto and Ebert 2010). In the context of Europe, and following the 2015 refugee crisis, a pan-European civic and activist mobilization effort created an unprecedented network of support for migrants with precarious legal status, giving a whole new meaning to being politically engaged and organizing while in transit.

Conclusion

By showing how various contexts at departure, as well as in the country of residence, structure social mobility for undocumented migrants, this chapter examined the complex ways in which undocumented migrants forge lives for themselves despite their precarious existence in different spheres. We explored the ways in which undocumented status shapes the potential for social mobility, focusing on how different demographic characteristics (age and gender, for example) may further facilitate or constrain it. The stories opening our chapter served a dual aim: on the one hand, to challenge and broaden the public imaginary regarding who is an undocumented migrant; and, on the other, to highlight the various social mobility pathways that undocumented migrants may follow.

As previous chapters illustrated, undocumented life and the specter of deportation may have detrimental effects on people's everyday experiences and their sense of belonging, and these quite often lead to a life at the margins. The challenges that many undocumented migrants face can be bleak and further emphasize how the demand for cheap labor comes at a high moral and human cost. While the labor options of undocumented migrants can often be limited and expose them to subpar conditions, unions, worker centers, and other community-based initiatives can alleviate some of these conditions. Additionally, different contexts of reception also play a role in the types of resources and experiences available to undocumented migrants and their families, thus shaping the opportunity structure.

The chapter outlined how these different factors impact undocumented migrants' social mobility: from pre-migration factors, such as class status in home countries and educational level, already setting migrants on certain paths, to immigration policy and policy environments in the host countries. As incorporation and social mobility are particularly place-bounded processes, we focused on these local contexts that facilitate or hinder them. National and local contexts, policy environment, civil society, and migrant communities in the host countries play an important role in structuring undocumented migrants' opportunities and vulnerability. As such, everyday practices of immigrants, those who work with them, and the communities they reside in, shape a range of experiences which fall between exclusion and belonging. Indeed, undocumented migrants in some localities are eligible for a driver's license and in-state tuition rates at public universities. And some communities may provide opportunities for undocumented migrants to participate more broadly in community life through voting in local elections and serving on local councils. In contrast, more restrictive areas can turn daily experiences, like driving a child to school in the morning, into a severe risk. The looming

Undocumented Status and Social Mobility

threat of apprehension and deportation can be more palpable in these areas. Additionally, the establishment of liminal or partial legal statuses can further complicate the concept of exclusion and inclusion for immigrants.

The chapter finally discussed schools as tools of integration and access to labor markets as a booster for social mobility. In particular, schools provide undocumented migrant students and their families access to community networks and other social services and, as such, serve as effective sites of inclusion. In turn, access to the labor market can be a significant constraint for migrant integration and social mobility. While meeting a demand created by global process, as we examined in previous chapters, undocumented migrants in particular are more often than not called to take up low-paid, unregistered, and even dangerous jobs.

6

Families and Children

On September 5, 2017, US General Attorney Jeff Sessions announced the termination of the Obama administration's Deferred Action for Childhood Arrivals program, an enforcement provision that had shielded nearly 800,000 young undocumented migrants from deportation since August of 2012. Despite notable limitations,[1] the program, commonly known as DACA, was widely popular among the American public. DACA was also, arguably, the most successful integrative immigration policy in the United States in over three decades (Ellis, Gonzales, and Rendón García 2018; Gonzales et al. 2018). While the future of DACA and the thousands of young adults who have benefited from it is still uncertain, the tranquility and opportunities it has brought for these young people cannot be overlooked and many continue to fight for DACA.

Over the course of DACA's first five years, its beneficiaries took giant leaps towards the American mainstream: through their work authorization cards they had obtained jobs that matched their education and degrees; they began building credit through acquiring credit cards and opening bank accounts; and through new opportunities and the passage of state policies they obtained health care and driver's licenses (Gonzales et al. 2018).

These new opportunities provided a significant segment of a broader population of 11 million undocumented migrants with the resources and motivation to pursue additional education and workforce development opportunities. It also allowed them the ability, through increased income, to more meaningfully contribute to their families' finances.

In Sessions's announcement, he outlined what would be a wind-down period of six months, giving Congress an opportunity to pass legislation to protect DACA beneficiaries. After facing legal challenges to ending the program[2] and growing pressures from a restrictionist base, the Trump administration opted for what they argued would be the "least disruptive" option available.[3]

For many advocates, the termination of the program represented an undoing of the gains made by young beneficiaries and their families in the five years of the program. What's more, they worried that it would put their lives in peril.

At a café in Chicago, Adelita Torres stared blankly at her biology textbook, covered neatly with a cut-out brown paper bag and masking tape to protect it from everyday wear.[4] In DACA's five years, she was able to make important investments in education and to achieve employment mobility as a result. Faced with the prospect of the program ending and losing her status, Adelita was at a loss how to think about her future.

Adelita and her family had moved to Chicago from her native country of Mexico in the winter of 1992 when she was six years old. They settled on the city's southwest side, initially renting out a living room from the cousin of their neighbor back in their hometown. Adelita's father had set out to Chicago more than two years previously on a tip from a friend that there was work in a pizza factory. He spent that time saving money and preparing for his family's arrival. But the cost to bring them from Mexico ate up most of his savings. When Adelita, her mother, and her sister joined him, they struggled financially. Her father's wages were not sufficient to meet the needs of the entire family. Eventually, Adelita's mother joined her father at the pizza factor, after he was able to secure a job for her. But this meant that the two girls spent most of their days not seeing their parents. Left in the care of the teenage daughter of their landlord, sometimes they did not see their parents for days on end.

After about 16 months, Adelita's parents were able to save up money and move to their own place – a studio apartment a couple neighborhoods over. However, over the years the family would continue to struggle. Making ends meet was difficult, especially on the wages paid at the factory jobs her parents undertook. And, after her brother was born, Adelita assumed major responsibilities for helping out at home – cooking, cleaning, ironing, vacuuming, and making sure her younger siblings got to school on time.

Adelita left high school during her junior year when an opportunity arose at a hair salon managed by her friend's aunt. The salon needed a front-desk receptionist who could schedule appointments and greet customers while also keeping the floors clean. Although she enjoyed her studies, she had become discouraged by her undocumented status and had lost hope of pursuing a fulfilling career.

When DACA was announced on June 15, 2012, Adelita was 26 years old. Returning to school had been an elusive prospect. She was still working at the salon. It was steady work – much better than the limited options her parents faced – and her hourly rate of US$12.25 was hard to beat anywhere else she could find work. But she yearned for more. She attended a DACA clinic at a nearby community center one evening after work and, over the course of two weeks, she enrolled in a GED program, gathered the documents she needed to prove eligibility, and purchased a money order for US$465 to process her application.

With her DACA status and a renewed sense of purpose, Adelita quickly moved through the GED program. She eventually passed her exam, but she did not stop there. She found a radiology program at the City Colleges of Chicago and, in two and-a-half years, she earned an Associate degree in Applied Sciences. With her degree and a little luck, she landed a job at a university hospital as a radiologic technologist, taking X-rays and CAT scans. The job paid Adelita more than US$35,000 a year. She felt as though she had finally turned her life around.

With the extra income Adelita earned, she was able to move her family into a new apartment, put a down payment on a used sedan, and help her parents with rent and bills. In a few short years, DACA had transformed not only Adelita's life but also that of her entire family. By the fall of 2016, she had enrolled in a four-year university course where she planned to study pre-med. She took classes by day and worked the second shift at the hospital. Adelita's schedule was exhausting, but she was driven to succeed.

But Sessions's DACA announcement had halted her plans. As Adelita stared at the textbook, tearing away at the paper that was covering it, her actions were an analogy of her present circumstances. "I really don't know what's going to happen to my family now," she said. "Everything seems to be coming undone. DACA allowed me to have a job, to have dreams. I've been able to do so much with it. Now I have no idea, I don't know what I'm going to do, what's going to happen to us."

Adelita is one of the more than 16 million people who live in mixed-status families in the United States. Despite her parents' best efforts to sustain the family through various menial factory jobs, they still struggled to make ends meet. Through DACA, Adelita was not only able to advance her own life but was also about to lift up her family in the process. Her story is a vivid example of the ways in which families with precarious immigration statuses rely heavily on each member to get by.

Liminal statuses like DACA can have substantial impacts on moving mixed-status families up or down the social mobility ladder. In Adelita's case, DACA opened opportunities for her to better her professional life while helping her family move into better housing. Despite these inroads towards better opportunities, the potential termination of DACA has put her and her family in limbo once again. To be sure, Adelita's precarious circumstances were not entirely suspended by Adelita's DACA status, yet it gave the family room to breathe. As Adelita's story illuminates, the lives of undocumented young people and their families are also shaped by "illegality." This chapter focuses on mixed-status families and the ways in which experiences of illegality complicate and reconfigure family life. In their efforts to lead meaningful lives, mixed-status families must continually negotiate often confusing and contradictory circumstances.

Family life within and across borders

Undocumented migration has long been perceived as an issue predominantly involving single adult migrants who traverse borders in search of work. But recent developments in migration patterns and family settlement have had profound effects on the composition of migrant populations in host countries. Undocumented migration is increasingly a family affair. Today, growing numbers of migrants live with family members in countries of destination throughout the developed world. In the United States alone, more than 16.6 million people live in mixed-status families in which at least one undocumented migrant resides, and nearly half of all undocumented migrants are parents of minors. Most undocumented adults are long-term stayers, with two-thirds having lived in the United States for more than ten years. More than four million US-born children have at least one undocumented parent. And, of an estimated population of 11 million undocumented migrants, more than two million have lived in the United States since childhood.

As families struggle to gain a foothold in their new countries, these children grow up in an unprecedented context of shrinking rights for noncitizens and heightened immigration enforcement. Today, while countries are compelled to protect children, they face increasing pressures to enforce immigration laws. For many receiving countries, the scales tip towards greater enforcement. These particular contours present new sets of challenges that require different frameworks for understanding.

Family separation

With the passage of the 1965 Hart–Celler Act, the United States solidified a priority to family members of US citizens or lawful permanent residents in matters of admission to the country. This focus on family reunification enabled certain migrating family members to reunite with those living in the United States by providing the means for the latter to sponsor their migrating family members and allowing them to adjust their immigration status in the United States.

But a series of developments, beginning in the late 1980s, involving the buildup of the border, the criminalization of immigrants, and increased interior enforcement measures began discouraging undocumented migration (Boehm 2016, 2017; Golash-Boza 2017). All told, these efforts have included building longer, taller, and thicker border fences, dramatically increasing the number of border patrol agents and making large investments in technology intended to apprehend border crossers (Massey, Durand, and Malone 2002). Coupled with a proliferation of restrictive immigration policies, these measures not only increased the number of undocumented migrants living internally, they have also made it more difficult for migrants living in the United States to bring their families from abroad (Abrego 2014; Boehm 2017). In addition, enforcement efforts spread throughout the interior – in neighborhoods, bus stations, public parks, and places of work – have ratcheted up levels of fear and anxiety in immigrants, narrowly restricting everyday life and putting undue strain on families.

In Europe, restrictionist immigration policies have reduced legal pathways to migration for lower-skilled migrants, particularly from the global South. As a result, migrant journeys to Europe have become costlier and more dangerous. In the face of limited legal avenues to migrate and to reunite with their family members, migrants are often pushed to undertake expensive and dangerous journeys. In 2015 and 2016 alone, more than a million people arrived by sea at the southeastern shores of the continent. Many of these migrants were unaccompanied children traveling to live with relatives in Europe or they were adult migrants joining their spouses. Their aim was to either join their families or, through family reunification, to settle and bring their families over as soon as possible. Alas, more than 12,000 people died trying to cross the Mediterranean sea between 2014 and 2016, many of them children.

Families that have a mix of undocumented members and citizens or legal residents – referred to in the literature as mixed-status families – and those that must live apart due to these restrictive immigration policies – transnational families – are exposed to the everyday experiences that characterize and define their precarious and uncertain status. Nevertheless, while nations possess the power to define formal membership, different aspects of people's everyday lives (local policies, institutions, community relationships) give meaning to forms of legal citizenship; being inside or outside politically drawn lines has important consequences.

As noted in chapter 5, where mixed-status families live can also have important implications for members' experiences of exclusion and belonging. Thus, as much as nations produce conditions of exclusion and precariousness, migrants' lived experience is fundamentally shaped by everyday interactions with people, institutions, and social structures. A mother's inability to obtain a driver's license, for example, can transform the act of driving a child to school to a deportable risk. And restricted access to public higher education forecloses future options for young people. At the other end of the spectrum, an undocumented migrant who is sponsored for legal residence is allowed unrestricted access to the city's health services, while voting in local elections or to serve on local school boards provides migrants opportunities to embed themselves and their families in the life of their community.

Transnational families

Separation is a common feature of migration, particularly for undocumented migrants. The act of leaving families in the country of origin in order to migrate abroad to work is part of a pattern that has persisted for hundreds of years. Prior to recent efforts to fortify national borders, when boundaries between countries were much more porous, migrants could make return trips home and then repeat the cycle all over again. However, investments in border walls and border security have made the act of crossing back and forth much more difficult, costly, and dangerous (De León 2015; Jusionyte 2018; Massey, Durand, and Malone 2002). With the disruption of circular migration, many migrants decide to settle their families in host countries where they lack formal legal status and the rights that typically accompany membership.

But the process of reuniting is often difficult and time consuming, taking months, years, or even decades and straining familial relationships. In Europe, family reunification has been one of the main channels of immigration into the continent. The right to family reunification of legally residing migrants is guaranteed by the European Council's Directive 2003/86/EC on the right to family reunification.[5] Nevertheless, rates of reunification vary greatly between member states. Some countries, such as the United Kingdom, require a sponsor to have sufficient income to sustain his or her family, while other member states impose a waiting period before a sponsor can even apply to reunite with his or her family. Since the beginning of the 2015 refugee crisis, family reunifications have been significantly delayed due to a bottleneck in the asylum system of many EU member states and a race to the bottom by EU member states to make family reunion harder to achieve (e.g., Denmark, Germany, and the Netherlands).

Today, transnational families migrate for many reasons – some for work, others because of war and conflict. They personify the layered, restrictive, and prohibitive nature of contemporary immigration (Abrego 2014; Boehm 2012; Dreby 2010). Although some families do migrate together, many do not have the resources or the right circumstances to do so. It is much more common for some family members to migrate first while others stay behind. Sometimes it's a parent or two, sometimes a son or daughter. This strategy allows family members to establish themselves in destination countries, thus smoothing subsequent migrations for family members. Also, many families simply do not have the resources to make the journey all together, even when their circumstances compel them to leave. As a result, families become bounded by separation.

Historically, patterns of labor migration have often been gendered, as men were more likely to migrate while women stayed home to care for children and family. It was often men who would establish themselves in receiving countries before sending for their families. But, beginning in the early 1990s, growing numbers of women began to migrate (Hondagneu-Sotelo 1994). New patterns of migration reconfigured family structures and dynamics. As the

Families and Children

journey became increasingly dangerous and marred by multifaceted forms of violence, it forced families to leave young children behind in the care of grandparents and other relatives (Dreby 2010; Menjívar, Abrego, and Schmalzbauer 2016). Those left behind often had to navigate dire situations of poverty, violence, conflict, and loneliness (Abrego 2009, 2014). For many families, separation was long and indefinite.

As sociologist Joanna Dreby poignantly underscores, for many immigrant parents living in destination countries, migration is a disadvantage (Dreby 2010). Sacrifices are experienced in numerous ways, including the types of job they take and the strain that migration places on families. Many parents forgo important time raising their own children. Meanwhile, they must contend with the difficult burden of providing for those with whom they migrated, while also tending to those they have left behind. Through remittances and sometimes gifts sent back home, migrants provide necessary income and meet material needs for their families that can range from day-to-day necessities like food to education costs for the children they left behind. Remittances can often be the sole source of income available to families in sending countries. But giving up large portions of their wages through remittances contributes to parents' stress and financial instability (2010).

According to sociologist Leisy Abrego, gender can play a critical role in remittance behavior. Female migrants, for example, confront multiple sources of vulnerability. Women often make sacrifices, forgoing their own safety to ensure that their family members back home have access to some form of financial stability. Sometimes this makes them vulnerable, even staying in abusive relationships so as to maximize available income to remit back to their families (Abrego 2009). Over time, these arrangements can lead to numerous difficulties.

Dreby argues that parents' sense of time differs from that experienced by their children (Dreby 2010). That is, for parents working long hours, time moves at a faster pace than for children waiting in their country of origin. The time spent getting settled and paying off debts often prolongs separations. Meanwhile, children are growing up out of view. As a result, time spent "divided by borders" also enlarges an emotional separation, sometimes punctuated by estrangement and resentment (Artico 2003; Horton 2009; Suárez-Orozco, Bang, and Kim 2010; Yarris 2017).

These contemporary patterns expose a hard reality. Parents often migrate in the hope of providing better economic opportunities for their children. But structural conditions create numerous strains on families (Abrego 2014). Parents who have left their families must find ways to cope with the emotional toll of living apart from their children while also maintaining their lives in the receiving country. Those parents who are undocumented find employment in situations that can expose them to the harsh realities of exploitation due to their lack of legal status. These everyday and long-term effects of illegality are not often seen by family members left behind in the country of origin, thus creating even greater emotional distances between parents and children. Despite all of these challenges, transnational families find ways to remain connected and maintain hope in the possibility of being reunited.

Mixed-status families

Due to policies and practices that privilege cheap and flexible labor and have made the act of border crossing increasingly difficult, a growing number of families are settling in receiving countries. More often than not, these families are not made up exclusively of undocumented members. Rather, they are a mix of statuses, what Leo Chavez once referred to as "binational families" (Chavez 1988). Typically, these families are headed by at least one undocumented parent who has children who are citizens or lawful permanent residents or even those who lack legal status but have grown up most of their lives in the host country (Abrego 2006; Dreby 2010; Gonzales 2011). These families illustrate the fragmented and capricious nature of immigration policies. Their presence complicates what we know about family roles, access to social services, and membership (López 2015).

Although family reunification has been a hallmark of US immigration policy, members of mixed-status families confront myriad difficulties in their attempts to adjust their immigration status.

Families and Children

Certainly, under current provisions for family sponsorship in the United States, those in families who possess full legal citizenship face easier and more direct pathways to sponsoring close relatives or other family members. But these processes are complicated in the European Union and the United Kingdom. And even, in some families in the United States, members' attempts to sponsor spouses can have devastating consequences (Gomberg-Muñoz 2017), leading sometimes to exposure, detention, and deportation of certain family members.

Anthropologist Ruth Gomberg-Muñoz argues that the immigration process is experienced differently depending on immigrants' countries of origin, gender, and sexuality. Moreover, because the criminal justice system is now so closely linked to the immigration system, certain immigrants are more likely to be ensnared in the former and, as a result, to face difficulty adjusting their immigration status. For example, the disproportionate targeting of undocumented migrants from Mexico and other Latin American countries for criminal violations places them in direct contact with law enforcement and jeopardizes their ability to adjust their status (Golash-Boza 2015a). What's more, unlawful entrants are more likely than visa overstayers to come from particular countries.

As a result, certain immigrants attempting to "become legal" are harshly exposed to the ways in which laws can work against family cohesion (Boehm 2016; Gomberg-Muñoz 2017; Menjívar, Abrego, and Schmalzbauer 2016). Although immigration policies can help keep migrant families together and reunite those who are left behind, these same laws can sometimes work against mixed-status families. Sometimes attempts to adjust the immigration status of a family member can lead to devastating consequences. Navigating a complex legal system is no easy feat, but many must also return to their countries of origin where they must engage in consular processing in order to adjust their immigration status. Once there, they may come to find out that they face barriers to returning, including multi-year or permanent bars to reentry. Others may land in detention and deportation proceedings through their attempts to adjust status.

Through restrictive laws and practices, host nations can also make life difficult for family members who possess legal status. Indeed, the lives of citizens and permanent residents are powerfully intertwined with those of their undocumented family members. A deportation, for example, can produce "suddenly single mothers" – women who, by the dint of their circumstances, find themselves having to care for their children and family members alone. These women often struggle to make ends meet in the wake of their husbands' deportation (Dreby 2015). It could also lead to the return migration of entire families who have made homes in the host country but who have difficulty meeting the emotional and financial toll of separation.

The toll of undocumented status

Migration alters and, at times fractures, family life. As we have mentioned in the previous section, it separates family members. In receiving countries, where mixed-status families carry out their everyday lives, undocumented status can enact a heavy toll on families. Undocumented migrant parents are limited to low-wage, clandestine employment. These grueling jobs often require them to spend long hours away from their families. What's more, these configurations reshape family life. Children not only take on extra work at home, their responsibilities carry over when their parents are unable to successfully navigate language barriers and systems they cannot understand. Apart from the emotional toll of living in the shadows, many adolescent and young adult children must also face everyday constraints of undocumented status.

In mixed-status families, immigration status and poverty are intimately connected. Due to this troubling mix of disadvantage, mixed-status families often live in high-poverty neighborhoods and their children attend low-achieving, under-resourced schools (Chavez 1991; Hondagneu-Sotelo 1994; Zlolniski 2006). As such, families face notable challenges moving from impoverished conditions to leading lives with financial and emotional security. They are much more likely to live in crowded housing, lack health care, and struggle financially (Capps et al. 2005). The average income of families with at least one undocumented parent is 40 percent lower than that of either native-born families or legal immigrant families.

In Europe, undocumented children often fall through the cracks, even though they possess rights and all EU member states are committed to their protection. While access to education and health is a universally recognized human right for children, there are some who are ejected from school due to their or their parents' undocumented status (UNICEF 2012). These problems often arise due to conflict between the obligation of states to protect children, on the one hand, and the growing trend towards immigration enforcement (Sigona and Hughes 2012).[6] Many of these children and their families live in constant fear of being deported. As these children are UK-born to undocumented parents, they have never lived outside the country, so they experience extreme anxiety in the face of their deportability. This often prevents them from thinking about their future and from standing up for their rights.

Undocumented status limits the labor-market participation of migrant parents. The limited range of options funnels migrant workers into employment in the informal economy, characterized by long hours, low wages, hazardous work environments, and scarce opportunities for job security and health benefits. These jobs also limit opportunities for self-direction and autonomy while taking parents away from their families for long periods of time. Taken together, these particular contours of undocumented labor have particularly negative effects on children and on parents' available time to interact with their children.

Mixed-status families are also more likely to lack health insurance and to avoid taking advantage of banking and other financial services (Fortuny, Capps, and Passel 2007). Parental fears of deportation can cause them to avoid institutions and institutional agents, thereby limiting their children's access to vital services, even when they are eligible, such as health care and quality child care (López 2015; Sigona and Hughes 2012; Vaquera, Aranda, and Gonzales 2014; Yoshikawa 2011). As a result, they often forgo services and programs that provide critical services and learning for their children. For example, they are less likely to apply for food stamp benefits, even when their children are eligible. And they often avoid applying to government agencies for their children's healthcare benefits.

While undocumented life is difficult for all in mixed-status families, children face particularly heavy burdens. Common stressors associated with mixed-status families can impact children's development across the life course (Ortega et al. 2009). Avoidance behavior, though a rational response to constrained circumstances, directs families away from a wide range of services that have traditionally benefited immigrant families (Fortuny, Capps, and Passel 2007; Menjívar and Abrego 2009; Yoshikawa and Kalil 2011). Especially when resources or services for children require forms of identification or proof of employment, undocumented parents tend to withdraw their participation. Predictably, this avoidance behavior removes children from important opportunities and services. In the long term, it hinders cognitive and motor development in children and negatively impacts their school outcomes (Bean et al. 2011). For undocumented children in mixed-status families, conflicting experiences of illegality and belonging start very early, as they often experience integration at school but also witness their parents' legal exclusion (Dreby 2015).

Family struggles are further exacerbated when children hold different immigration statuses and thus have differential access to services (Abrego 2016). While some members of a mixed-status family may benefit from a wider array of services because of their immigration status, others in the same family may be locked out of access to these same services, creating conditions for competition and resentment (2016).

These contours of undocumented life compel family members to take on multiple roles within the family unit. Thus children in mixed-status families are often tasked with an inordinate amount of household responsibilities. For example, depending on children's own immigration status and sometimes their age, they often serve as cultural, social, and political brokers for parents and other relatives (Buriel and De Ment 1997; Katz 2014; Street, Jones-Correa, and Zepeda-Millán 2017). Children take on roles as translators and interpreters for the families, providing bridges between their parents and community institutions, such as schools, utility companies, and social service agencies. This is particularly common for children of the 1.5- and second generation since they have access to US schooling and other services which may provide them with socio-cultural knowledge that can assist them and their families.

In addition, children are often asked to take on a heavy share of responsibilities for the household. These duties often reflect gender norms, with daughters assuming caretaking responsibilities for siblings and other household chores, including cooking, cleaning, and ironing. Sons, on the other hand, assume responsibilities outside of the household, often entering the workforce at early ages. For many of these young males, entry into the world of work is made possible through family business or ethnic niches where other family members participate. Even when both male and female children enter the workforce, the types of jobs available to them, through family business or other networks, are often gendered (e.g., boys taking jobs in construction and landscaping and girls finding work in housekeeping and child care).

As much as children are asked to pitch in, heightened responsibilities within the family can upset power dynamics and create tensions within the household. The relative power that children wield outside the home can disrupt the authority of parents and create an imbalance of power within the family (Menjívar et al. 2016; Portes and Rumbaut 2001; Schmalzbauer 2014). With parents limited by language and low educational attainment, children can more easily navigate institutions and society more generally. These dynamics can lead to role reversal, with children holding power over their parents in certain situations. Children can also grow tired of the additional roles and related stress and begin to resent their parents for the responsibilities they must bear as translators and socio-cultural brokers.

Despite these challenges, mixed-status families find ways to overcome their precariousness. Children can gain important skills and early awareness of pertinent issues which can help them better navigate difficult circumstances and become more civically involved (Nicholls 2013; Street et al. 2017). Skills learned as children can be useful tools that not only affect their family members but also themselves.

Schooling and the transition to illegality

Adolescence is widely understood as a period during which young people begin to develop unique identities, take on adult responsibilities, and grow increasingly independent of parents by leaving the home, pursuing higher education, and starting careers (Erikson 1963). Yet, for undocumented youth, the dawning of adolescence has been likened to a waking nightmare (Gonzales and Chavez 2012), in which youth begin to encounter legal barriers associated with their undocumented status and come to terms with how these will impact their future prospects.

While adult migrant parents are absorbed into the United States primarily through the labor market, their children are integrated into the country's social and cultural fabric through K-12 schools (Gleeson and Gonzales 2012). Schools provide immigrant children and the children of immigrants the opportunities to learn the language, customs, and culture of their new country (Suárez-Orozco, Suárez-Orozco, and Todorova 2008). They also allow them to integrate into a peer group that will experience common milestones together (Rumbaut 1997). As such, primary and secondary schools are a key socializing force (Gonzales 2010b; Gonzales, Heredia, and Negrón-Gonzales 2015).

Because they are constitutionally protected to attend American schools, undocumented migrant children grow up alongside American-born and citizen peers. In school, they enjoy the same baseline opportunities afforded to their peers. But while laws treat children and adults differently, they do not allow for the continuity of children becoming adults. As undocumented migrant youth reach adolescence and seek to pursue adult lives, they encounter a greater number of barriers associated with their immigration status. At a time when their friends begin to take after-school jobs, obtain driver's licenses, and go out to the polls to vote, they find themselves unable to participate (Abrego 2006, 2011; Gonzales

2011; Suárez-Orozco, Yoshikawa, Teranishi and Suárez-Orozco 2011; Yoshikawa, Suárez-Orozco, and Gonzales 2017). Instead, their undocumented status blocks their attempts to participate in adult life. In addition, at any time they can be deported back to their countries of birth.

Scholarship on the transition to adulthood has provided important frameworks from which to understand this developmental period as young people begin to exit roles of dependence and take on a greater number of adult roles (Arnett 2000; Berlin, Furstenberg, and Waters 2010; Côté and Allahar 1996; Erikson 1963; Osgood et al. 2005). Sociologist Roberto G. Gonzales turns this concept on its head by arguing that as undocumented children begin to come of age and make adolescent and adult changes, they transition to lives defined and narrowly circumscribed by their undocumented status (Gonzales 2011, 2016). Just as their peers ready themselves to obtain driver's licenses, take after-school jobs, and pursue college, undocumented adolescents begin to recognize the limitations of their childhood inclusion and struggle to make sense of their future lives (Gonzales 2011).

This chain of events depresses motivations and hampers the pursuit of education and employment. Additionally, as undocumented migrant youth begin to experience blocked access, responsibilities to their families increase. This prompts many to seek clandestine employment in the informal economy. The combination of limited financial resources, family circumstances, and steep educational costs force many undocumented adolescents to enter the underground workforce. However, many find themselves ill prepared for the demands and pace of the low-wage workforce, given their lack of work experience. Many also struggle with the difficulties associated with their new realities. Having to toil in low-wage work that is often physically demanding and dangerous is a difficult pill to swallow, especially for young people who grew up believing that their educational attainment and English proficiency would provide them opportunities not afforded to their family members.

Over time, their everyday lives become increasingly framed by their undocumented status. Work life drives them further out to the margins. And engaging in mundane acts like working and driving increases the risk of police encounters and deportation (Suárez-Orozco et al. 2011; Yoshikawa et al. 2017). Ultimately, they learn to avoid immigration authorities and comply strictly with traffic laws. They develop vigilant responses to their vulnerability, always looking over their shoulders in fear of the police. In addition, they come to have a greater awareness of their identities as undocumented migrants and develop behaviors to manage stigma (Abrego 2011; Castro-Salazar and Bagley 2010; Gonzales 2016) These avoidance behaviors include being vigilant towards strangers, avoiding potentially threatening situations, and concealing their status from authorities, employers, and even friends (Gonzales and Chavez 2012). Without legal opportunities to build more stable adult lives, many undocumented young people learn to cope with their uncertain futures by lowering their life expectations and divesting themselves of long-term plans (Gonzales et al. 2013).

This process has been identified as a "transition to illegality," a progression involving a complete retooling of self-perceptions and aspirations to match the narrowing of everyday life (Gonzales 2011). For the vast majority of undocumented youth, this involves opting out of educational pursuits. A minority of youth who have access to social and economic support manage to pursue higher education and so delay entry into the low-wage labor market (Abrego 2008; Cebulko and Silver 2016). However, without changes to their immigration status, even these young people eventually run out of educational options and join their aforementioned peers. Consequently, regardless of when they begin the transition to illegality, by the time they reach their late twenties, undocumented young people enter the shadows of American society, navigating daily challenges, exclusions, and fears that resemble the lives of their undocumented parents more than their citizen peers (Gonzales 2016).

By their late twenties, undocumented young people find themselves coming to terms with lives of limitation. Over time, the constraints of undocumented adult life take their toll on young people's minds and bodies, and they learn to accept that "their immigration status – not their dreams – would shape most of their future plans" (Gonzales 2016: 198). Undocumented status becomes a "master status," overwhelming all other achievements and identities. Completing this transition to illegality is arguably most difficult for young people with high educational achievement because of early opportunities that shielded them from some of the more negative aspects of undocumented status. But they, too, ultimately join their parents and other undocumented migrant adults in low-wage jobs and lives circumscribed by their undocumented status.

Unaccompanied migrant youth

Not all young people migrate with families or with people they know. Some must face the challenge of clandestine migration on their own, risking their lives along the way in the hope of finding safety in a new land. During the summer of 2014, about 60,000 children under the age of eighteen were apprehended at the US–Mexico border and turned over to federal custody. The United States classifies these children as "unaccompanied alien children" (UAC) since they have no legal immigration status and no parent or legal guardian to care for them in the United States (Chen and Gill 2015; Terrio 2015). Importantly, given their lack of legal status, these minors are placed in deportation proceedings and forced to navigate a complex and bureaucratic legal system (Heidbrink 2014). Many of them are eligible for asylum or other forms of legal status adjustments. Although about 78 percent of those who do file immigration cases receive some form of immigration relief, a staggering majority – about 97 percent – do not receive a simultaneous grant of legal immigration status. This means that they remain unauthorized, which continues to leave them vulnerable.

Most of these young people continue to come from the Northern Triangle, an area in Central American made up of El Salvador, Guatemala, and Honduras. These children and young people are fleeing multiple forms of violence, including gang violence,

Families and Children

war-like conditions, gender-based violence, and state-sanctioned violence. They cross multiple borders to make it to the US–Mexico border where many are apprehended and put in detention centers. The age range of unaccompanied children coming through the US–Mexico border has been falling, with children as young as eight arriving on their own.

Unlike the undocumented migrant youth who come to the United States at a young age to live with family members, unaccompanied adolescents join the ranks of the estimated 2.4 million undocumented migrants under thirty who arrived when they were fifteen or older (Passel and Lopez 2012). Given their older age, their incorporation into a new country is not as smooth as that of younger arrivals (Canizales 2015, 2018; Diaz-Strong and Ybarra 2016). They are arriving at a critical point in their adolescence, with varying levels of education and in legal limbo. Not only must these adolescents adapt to life in a new land but also wait for their asylum cases to be resolved or else face deportation.

At the peak of Europe's so-called "refugee crisis," in 2015, a record number of unaccompanied migrant children reached the

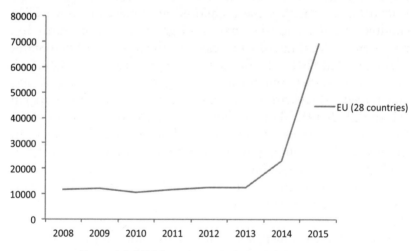

Figure 6.1 UASC applications in the EU, 2008–2015.

Source: Eurostat 2018; Elaboration: Becoming Adult, https://becomingadultproject. files.wordpress.com/2017/12/ba-brief-5-low-res.pdf

European Union and lodged formal asylum claims. Countries such as Sweden, received in one year more than three times the European Union's average number of asylum applications. This unusually high incidence of migrating children placed a significant strain on child protection services in Europe and left many young people poorly supported during the status determination process. Thousands of unaccompanied minors arriving in Greece and Italy successively went "missing" (Humphris and Sigona 2016; Sigona, Chase, and Humphris 2017a, 2017b). Some eventually resurfaced elsewhere in Europe and lodged asylum claims there; others, fearing deportation to their country of origin, opted to live in the shadows.

Invisibility to the authorities, especially for unaccompanied minors without support networks, makes these young people especially vulnerable to exploitation and destitution. Research led by Elaine Chase and Nando Sigona on the transition to adulthood of former unaccompanied minors shows that young people who arrive as unaccompanied migrant children and then make the transition to adulthood with or without precarious legal status are largely missing in the European Union's policy agenda (Chase 2017; Sigona et al. 2017b). Policies governing the lives of these young people are highly restrictive in terms of possible outcomes, assuming that a return to certain countries of origin is a viable, realistic, and even preferable option. Chase's and Sigona's work demonstrates that this assumption is highly problematic and that the current repertoire of policy responses undermines young people's well-being and rights, while simultaneously proving ineffective. The vacuum of protection and the uncertainty concerning the ultimate outcome of one's application is illustrated by Kamran, a young Afghani boy who managed to stay hidden for three years before eventually gaining legal status.

> I was refused, and after I decided . . . they asked me to go signing for the reporting to the immigration office, which I didn't go. I left it, I left the house, I left everything. I ran away. For three years. I, just recently, I did a fresh claim and I got it. I got my residence permit . . . I was living with friends, one night here, one night there . . . it was rough.

It was like feeling back in Afghanistan. And I had a problem with eczema on my hands, but I was scared to go to the hospital.

Conclusion

This chapter explored family migration both as a process and as a phenomenon that changes over time, mostly – but not only – due to increasingly restrictive immigration policies and legislative regimes in the United States and in Europe. Faced with ever-mounting barriers to legal family reunification, and also with the escalating dangers and costs associated with migratory journeys, families must increasingly bear the toll of living separately. This chapter therefore focused on transnational and mixed-status families and the ways in which experiences of illegality complicate and reconfigure family life. We explored how migration affects and, at times fractures, family life by separating family members, creating not only transnational families but also mixed-status families who strive to lead meaningful lives, continually negotiating confusing, often contradictory, circumstances and shifting regulations.

The chapter has defined the mixed-status family and documented the extent of this phenomenon in the United States. We briefly illustrated ways in which having an undocumented relative may impact on the life of the whole family, especially on children. Adelita's trajectory illustrated the kind of opportunities offered to undocumented migrant young people by the DACA program in the United States to transform not only their own lives but also that of their families. Her story is also a vivid example of the ways in which families with precarious immigration statuses rely heavily on each member to get by. Consequently, the announcement by the Trump administration of the end of the DACA program has had a detrimental ripple effect on people's livelihoods and prospects, putting their lives in limbo once again.

We subsequently unpacked the role that family increasingly plays in migration journeys as opposed to the widespread view of international migration as a single adult, often male, endeavor. Although some families migrate together, most do not possess the resources or the right circumstances to do so. Hence usually some members of a family migrate first while others stay behind. Contemporary migration patterns expose a hard reality: while parents often migrate in hopes of providing better economic opportunities for their children, the structural conditions they experience create numerous strains on their families. We documented the available channels for family reunification on both sides of the Atlantic and illustrated the increasingly narrowing opportunities for legal migration. Coupled with restrictive immigration policies in host countries and tougher border controls, migrants are pushed into costlier and more dangerous journeys in order to reunite with their families or end up staying separated for even longer periods. Time spent apart, "divided by borders," has detrimental effects on family relations and especially on children growing up as it widens the emotional separation and leads to estrangement and resentment.

The chapter has paid particular attention to children and young people, and the toll that being undocumented or part of a mixed-status family has on them: fear of deportation, limited presence of their parents because they have to take up grueling jobs, poverty, lack of health insurance, attending low-achieving and under-resourced schools are only the more straightforward challenges they face. Living with undocumented status or being part of mixed-status families often forces children to take on inordinate amounts of household responsibilities that their parents are unable to because of language barriers or because of their lack of status. For example, depending on children's own immigration status and sometimes age, they often serve as cultural, social, and political brokers for parents, siblings, and other relatives. Finally, we examined what has been termed a "transition to illegality" which many undocumented adolescents experience as they eventually come to terms with lives of limitation. They tend to be faced with their undocumented status as they enter adolescence and question their sense of belonging. This status and their increasing awareness of it usually crashes their dreams and limits their job opportunities and prospects for further education and careers.

7

Challenging Exclusion

On May 17, 2010 five students, dressed in graduation caps and gowns, held a sit-in at the Tucson, Arizona office of US Senator John McCain. One of these students was 25-year-old Lizbeth Mateo. "I've been organizing for years, she told the *New York Times*. "And a lot of my friends have become frustrated and lost hope. We don't have any more time to be waiting. I really believe this year we can make it happen" (Preston 2010).

This action was part of a series of escalated actions across the country by undocumented migrant young people – sit-ins, hunger strikes, and acts of civil disobedience – to advocate for a concrete path to citizenship for undocumented migrants who came to the United States at young ages. In particular, this group has advocated for the passage of the Development, Relief, and Education for Alien Minors Act, commonly known as the DREAM Act, first introduced in 2001 by Illinois Senator Dick Durbin and Utah Senator Orrin Hatch. Over time, their actions have become one of the most impressive social movements of our time. And since those actions in 2010, their goals have expanded to include advocacy for their parents, other family members, and the entire population of undocumented migrants living in the United States.

Historically, immigrants and other marginalized groups who have been denied access to resources and political power have found extra-institutional avenues to assert claims and seek change. They have used street protests, acts of civil disobedience, rallies, and marches (Voss and Bloemraad 2011). On a smaller scale, undocumented migrant youth have held bake sales, car washes, and banquets to raise money for scholarships. At the local level, they have staged mock graduations to bring awareness to their plight. And they have raised the stakes by engaging in hunger strikes, sit-ins, and other acts of civil disobedience. These efforts, both local and national, have allowed them to consolidate their power and made them a potent political force (Gonzales 2008; Nicholls 2013; Rincón 2008; Seif 2004; Zimmerman 2012). While their actions may seem like a big risk, they view their efforts as a way to stake a claim on a political world that excludes those without legal citizenship.

Drawing on previous immigrant rights struggles – the Sanctuary movement, Justice for Janitors, and the Southern California drywall strike – this new immigrant movement, led primarily by young people directly impacted by US immigration policy, has brought a new sense of urgency to the issue of immigration reform. It has also allowed these young people to expand their networks while advocating for themselves.

Along the way, these young people, commonly known as DREAMers, have compelled business leaders, university presidents, and elected officials to support their efforts. They have swayed public opinion on the DREAM Act and broader immigration reform efforts. They have been instrumental in the successful passage of legislation supporting in-state tuition and driver's licenses for undocumented migrants. And in 2012, more than ten years after the introduction of the DREAM Act, their pressure on President Obama ultimately led to the implementation of the Deferred Action for Childhood Arrivals (DACA) program, a change in the administration's enforcement policy temporarily deferring deportation from the United States for eligible undocumented youth and young adults and offering these young people temporary social security numbers and two-year work permits.

Despite DACA's success, the United States continued to deport undocumented migrants in record numbers. In 2013, Lizbeth Mateo made headlines again when she – along with Lulú Martínez and Marcos Saavedra – crossed the US–Mexico border in an effort to bring attention to the effects of deportations on families. From Mexico, Lizbeth wrote, "undocumented immigrants run the risk of being taken from their home, no matter where we are ... It's time to take away the power deportation has over us" (Mateo 2013).

In the European context, the best-known example of mobilization by undocumented migrants began in France in 1996, when 324 undocumented migrants – mostly from Mali and Senegal – occupied the Saint Ambroise and Saint Bernard churches in Paris. While church occupations were at times violently evicted by the riot police, following a 50-day hunger strike, the *Sans Papiers* (literally, those without papers) movement was born and grew to be one of the longest-standing movements of undocumented migrants in Europe. Protesting against deportations and demanding the regularization of their status, the *Sans Papiers* movement was one of the first self-organized political movements led by undocumented migrants in France and Europe.

The *Sans Papiers* rejected the juridical categorization of people into dichotomous categories of citizens and noncitizens and sought to expose the state's practices of illegalization (Badiou 2009). *Sans Papiers* managed to make visible the struggle of undocumented migrants not simply as one for the recognition of their status and their entitlement to certain basic rights but also as one of long-term residents and workers who had been residing in the country for years, even decades, paying taxes and participating in the polity. Their claims on membership reanimated a notion of citizenship as a "collective practice" (Balibar 1996). In doing so, their movement created new spaces of politics and representation, bringing to the center those who were previously at the margins (Rodríguez 2004). It also sowed the seeds for migrant justice struggles today (Nail 2015: 110).

In this final chapter, we move beyond the largely structural discussion of illegality to examine the ways in which undocumented migrants exercise agency in asserting rights claims and their own belonging (Barrett and Sigona 2014; McNevin 2013; Nicholls 2013; Nyers and Rygiel 2012). We include an overview of existing national discourses on undocumented migration, the "politics of belonging" that undergird migrants' strategies and the various responses they encounter.

As we have discussed throughout this book, countries across the globe are increasingly tightening their borders. But borders are no longer confined to the territorial boundaries of the state. Sophisticated border surveillance technology, more stringent immigration policies, the expansion of enforceable territory, and identification checks performed by institutional actors are just some of the measures put in place over the last two to three decades that have drawn increasingly bold lines between and within nations. The focus on expanding border enforcement has been justified through concerns with international terrorism and a resistance to the pressures of globalization. These measures have caused human mobility to be highly regulated and closely monitored, while access to permanent residence and citizenship is more elusive and internally differentiated. For migrants who journey through deserts, rivers, and the sea to flee poverty, violence, and other circumstances, such stringent exterior and interior immigration policies can make them vulnerable to the growing inequality experienced by marginalized people.

Widening inequalities, both globally and within nation-states, have given rise to new forms of resistance from undocumented migrants and their allies. These instances of contestation provide opportunities to challenge increasing hostility towards migrants and minorities and to open political spaces where new and different articulations of membership can be introduced.

Redefining membership

As mentioned in chapter 5, undocumented migrants create bonds and form permanent homes despite their lack of legal citizenship. Through these acts of incorporation, undocumented migrants make claims to different forms of citizenship and contest the overly simplified "legal–illegal" binary in defining membership.

To understand these processes, it is useful to move away from T. H. Marshall's (1950) classic definition of citizenship as a status granted to those that are full members of a community and consider instead contemporary critical work on citizenship that theorizes citizenship as practice and performance rather than as a static immigration status (Isin 2002, 2008). This shift in perspective away from a static view of citizenship enables us to consider the practices of contestation of undocumented migrants as ways of claiming a stake in society and enacting citizenship before it is actually formally granted.

Through the mobilization and self-organization efforts of undocumented migrants and their supporters, conventional notions of citizenship and membership have been redefined and reconfigured. By crafting a place for themselves in the sociopolitical domain, undocumented migrants make "rights claims" and build their legal consciousness, thereby producing their cultural citizenship (Abrego 2011; Coll 2010; Nicholls 2016; Rosaldo 1994; Stephen 2003). As members of local communities become aware of the specific ways in which legal contexts shape their everyday worlds, they develop an awareness of their sociopolitical positioning in society and hence find ways to push these legal boundaries (Silbey 2005). The development of a legal and political consciousness has been facilitated through various avenues, including grassroots organizing campaigns, know-your-rights workshops, unionizing efforts, and English as a Second Language classes (Gleeson 2010; Nicholls 2016; Stephen 2003; Stuesse 2016; Voss and Bloemraad 2011).

Making rights claims for undocumented migrants can be a tenuous process, given the threat of deportability and the vulnerability that undocumented migrants experience (De Genova 2002; Gleeson 2010). While mobilizing undocumented migrants can be challenging, undocumented migrants and their supporters have created channels through which undocumented migrants can make strides within the boundaries of "illegality" and propel policy makers and lawmakers to acknowledge some of their claims for workers' rights, healthcare rights, and even the creation and extension of temporary forms of legal statuses (Coutin 2000; Heredia 2016; Marrow 2011; Menjívar 2006; Nicholls 2016; Stuesse 2016; Unzueta and Seif 2014). This civic action contributes to undocumented migrants' "cultural citizenship," or the contributions undocumented migrants make in their everyday lives to build a legitimate space for themselves within the nation-state (Flores and Benmayor 1997; Gálvez 2013; Rosaldo 1994; Stephen 2003). Through tangible economic, social, and cultural contributions undocumented migrants and their families are able to be make claims for their inclusion and validate their civic existence at the local, state, and federal levels.

Citizenship itself is not an evenly distributed status. Gender, class, race, and ethnic background considerably shape the ways in which immigrants enact citizenship and experience it. Therefore, considering inclusion and exclusion in reference to the state and to the binary categories of "legal" and "illegal" severely limits our understanding of the more complex ways in which migrants practice belonging and membership on the ground. These practices have been explored by many scholars who call for a decoupling of citizenship and the state in favor of a global or post-national citizenship (Bosniak 2006). Others have focused on more informal expressions of membership as these result from everyday encounters, relations, and interactions in both urban spaces and in migrant camps. Finally, researchers have also detailed the ways in which political struggles and other forms of political participation could form the basis for active forms of citizenship (Isin 2008).

In other words, "[m]igration scholarship has shown that the identities, rights and practices associated with being a citizen can be at odds with formal citizenship status, and that the entitlements conferred by citizenship can be obtained through claims to membership that supersede the nation state" (Barrett and Sigona 2014: 2). This disrupts the binaries of inclusion and exclusion, citizenship and noncitizenship, and insiders and outsiders.

Along with scholars who have documented the politicization of undocumented migrants, Anne McNevin (2013) finds in the nature of migrants' political claims the potential for resistance against migration restrictions (2013: 183). She uses the notion of "ambivalence" to provide a critique of both perspectives that theorize the autonomy of migration on the one hand and the work of Agamben (1998) on sovereign power and "bare life" on the other.

Ambivalence, McNevin argues, is a key characteristic in undocumented migrants' struggles and may indeed have transformative power over political subjectivities. Ambivalence towards formal citizenship and human rights, rather than a disadvantage, "may, *in and of itself*, inaugurate new kinds of political relations across the terrain of human mobility and border control" (McNevin 2013: 185).

Roberto G. Gonzales and Nando Sigona (2017) borrow from Linda Bosniak's conception of citizenship to expand on the "soft borders of citizenship," inviting us to rethink ideas and processes of inclusion and exclusion and how these relate to legality and illegality. They invite us to pay attention to the gray spaces between exclusion and inclusion in which belonging is "conditional, partial, temporary, and revocable" (2017: 8).

Similarly, Sarah Spencer (2016) examines the uneven geographies of entitlement of undocumented children across the European Union. She claims that this "postcode lottery" results from a generalized politics of deservingness that constructs undocumented migrant children as both "illegal" and vulnerable. This contradiction is reflected in the competing policy objectives of immigration control and child welfare and is often situated between different tiers of government, i.e., national vs subnational levels. She finds that, despite significant variations between member states in the access to services for undocumented children, ethical standards and requirements complying with welfare and human rights principles are central in the making of decisions regarding who receives services (2016: 1624).

Marching for freedom

On July 4, 2012, nearly two hundred undocumented migrants from more than twenty-five nationalities arrived in Strasbourg, France. This gathering marked the culmination of the "march for freedom" that had started a few weeks earlier in Paris, France. The march brought together migrants and activists from a number of EU member states, in particular France, Belgium, Germany, and Italy. In one month, the marchers crossed seven international borders, traversed 1,900 km and staged protests in twenty cities. Sociologist Thomas Swerts recalls a conversation with one of the marchers in Strasburg: "I traversed more than 400 km by foot, can you believe that? At the border between Switzerland and Italy, we stood face to face with the police, but the police did not intervene while we crossed the border! Isn't it fantastic that the police is making way for us, instead of vice versa?" (Swerts 2017).

The March for Freedom was, Swerts explains "a collective experiment in creating new forms of political belonging, participation and membership that supersede national borders" (Swerts 2017). The freedom marchers asserted that their mobilization was not directed at individual EU member states but was instead constructed as an EU-wide intervention. With their invocation of rights to the EU, the marchers were attempting to create the European Union as a political space not just for EU citizens but also for undocumented migrants. Their ultimate claim, therefore, was to be recognized as members of the European Union family, a concept that exercises a different degree of appeal within the European Union itself. Through the contestation of the borders and boundaries that define national membership, these non-citizens attempt to transform the meaning of EU citizenship from below, while revealing its weakness as a concept still lacking firm grounding. Hence, Swerts adds: "The case of the European march simultaneously points to the transformative potential of non-citizen political subjectivities while recognizing the limits and constraints placed upon them by existing citizenship paradigms."

The challenge for undocumented migrants to mobilize beyond national boundaries in the European Union was previously observed by Virginie Guiraudon (2001) in the late 1990s when attempts to mobilize EU-wide solidarity with the *Sans Papiers* movement failed to reach any long-lasting result. Guiraundon's analysis points to a series of structural gaps in EU policy architecture, as well as to a more cultural and political one. She explains: "Migrants face a more subtle challenge: that of being accepted as full members of the polity implicitly forces them to act as natives rather than cosmopolitans in a way that would not apply, in contrast, to native – regional or ethnic – groups making self-determination claims through international organizations" (2001: 180).

Space and scale are also consequential for the multiple exclusions and struggles noncitizens face (Maestri and Hughes 2017: 626). The border, the city, the migrant camp, the detention center are all critical sites where marginalization and exclusion take shape and materialize. At the same time, these sites create new spaces of encounter between migrants, authorities, and activists. They can also give rise to new processes of political subjectification and solidarity. These spaces also showcase the ways that struggles are articulated by the migrants themselves, their allies, the authorities, and other key players. Essentially, these articulations serve to frame both the way movement leaders organize and the state's response.

Seeking refuge

On Sunday, April 29, 2018 some 150 migrants reached the US–Mexico border. These migrants were part of an immigrant caravan that originally included more than 1,300 migrants, primarily from Central America (Guatemala, El Salvador, and Honduras) who traveled from the border of Guatemala and Mexico through Mexico seeking asylum (Semple and Jordan 2018). Those who made it to the US–Mexico border were greeted by several American advocacy groups and other supporters. One of these was Heather Cronk, co-director of Showing Up for Racial Justice, a national advocacy organization. Cronk explained why she was there: "For us, this is all about who we are as a country. This is an existential moment. This is a spiritual moment. I want it to be true that when we say 'Liberty and justice for all' we mean it" (Semple and Jordan 2018).

While only a portion of the original caravan members made it to Tijuana, their journey was well documented by numerous national and international media outlets. The caravan itself caused many immigration restrictionists to raise questions of border security, violence, and safety. Most of the migrants who made it to the US–Mexico border sought safety from the pervasive violence they had

Challenging Exclusion

experienced in their countries. Bayron Cardona Castillo, who traveled with his two-year-old daughter from Honduras, explained that all he wanted was to live in "peace and tranquility" (Semple 2018). While most of the migrants were merely seeking a safe refuge, the fact that their journey north took such a public stage compelled those in the United States and in Mexico to grapple with questions regarding asylum, refugee status, and their immigration policies.

There is burgeoning academic literature on migrant and refugee camps that examines them as critical sites for the study of exclusion, agency, and politicization. Critical border and migration studies have followed two prominent streams of thought regarding the understanding of refugee camps and camp-like spaces. On the one hand, scholars have elaborated theoretically and empirically on the highly influential work of Giorgio Agamben on exception, bare life, and sovereignty. On the other hand, drawing from a Marxist tradition, scholars have also focused on the autonomy of migration as a process. They stress the transformative power of human mobility and the complexity of endogenous and exogenous factors that shape it. A great deal of scholarly thinking has theorized the migrant camp and camp-like spaces as spaces of exception and de-politicization that produce abject subjects (Isin and Rygiel 2007), reduced to bare life (Darling 2009; Diken 2004; Edkins 2000; Papastergiadis 2006; Redfield 2005). As Walters points out, "Agamben focuses on what is done to migrants rather than on what migrants can actually do" (2008: 188).

The autonomy of migration, in contrast, focuses on migrant agency, that is, on their actions, and views migration as a force that precedes and challenges border controls (De Genova 2009; Mezzadra 2004; Mezzadra and Neilson 2013; Mitropoulos and Neilson 2006; Redclift 2013). This perspective understands migration policies and border controls as attempts by states to control at a global scale the mobility of the global poor (see Jones 2016). The "battle for the border" (Rodriguez 1996: 2), therefore, is about subjecting people to the vulnerabilities of the labor market (Rygiel 2011: 3). Similarly, other scholars argue that the camp, viewed from below (Papadopoulos 2008), allows considerations of it as a social and political space (Ramadan 2008, 2013) of struggles in everyday encounters and new relationships. These considerations also give rise to questions surrounding the politics of citizenship (Sigona 2015) and new political subjects (Rygiel 2011), enabling new emplaced forms of participation and membership to emerge, what Sigona has called "campzenship" (2015).

Drawing on his ethnographic work in Roma settlements in Italy and trying to make sense of the ambiguous position of camp residents vis-à-vis the state and their capacity to access welfare resources and "soft rights," Sigona develops the concept of 'campzenship' to capture the situated form of membership produced by the camp, in which residents are able to negotiate some form of recognition and political existence while experiencing spatial marginalization and status insecurity.

Rygiel (2011) argues that the migrant camp must be viewed as a social and political space that gives rise not only to border struggles but also to new enactments of citizenship by noncitizens. Reflecting on the dismantlement of the Calais Jungle in France in 2015, Rygiel shows how the camp became a symbol of resistance against increasingly restrictive border controls. It is through migrants' physical and bodily presence there, occupying space and being visible and resourceful in making communities and forging relations, that they become political.

Drawing on her ethnography with North African undocumented families squatting in houses in Rome in Italy, Rosa Parisi (2017) shows how a regime of citizenship, produced by governmental policies of a neoliberal state, is challenged by the struggles of migrant women who dispute and open the boundaries of the political body. She reveals the productive nature of these practices of citizenship and their ability to question the subordination produced by borders on their bodies. Parisi's attention, however, is not focused on the immigration status of the migrants but on their actions. This lens enables her to show the emergence of solidarity with "failed citizens," in her case impoverished Italian families in casual and poorly paid jobs who rely on state subsidies that barely cover rent and utilities. Through their personal experience with poverty, these families are able to join the migrant families in their claim for the right to adequate housing in Rome.

Identity and migrant politicization

As we have discussed in previous chapters, undocumented status impacts the everyday experiences of undocumented migrants and greatly shapes their identity. This process, in turn, defines the possibilities for political participation, civic engagement, and activism of undocumented migrants and especially undocumented youth. Identity, in this context, is characterized by both legal and social contradictions (Negrón-Gonzales 2014). The condition of "illegality" criminalizes undocumented migrants who are then subject to disparate treatment compared to those who have obtained citizenship or status constructed in the state in which they reside (De Genova 2002; Negrón-Gonzales 2014).

The sociopolitical processes of their own illegalization and racialization mark "both distinct forms of regulation and exclusion" (Negrón-Gonzales 2014: 259) and explain how experiences shape their participation and engagement in certain political activities. There is a constant negotiation within both the public and private sphere. Anthropologist Genevieve Negrón-Gonzales builds a framework for the understanding of the ways in which undocumented youth 'recast their insider status' (2014: 261) produced by their daily social interactions within society and the legal exclusions resulting from their lack of status. The youth re-articulate their exclusion as inclusion and thus perpetuate their political participation and engagement at a public level. As she notes, "(re)-articulation allows us to understand how political subjects make sense of lived contradictions and how dominant cultural practices can empower marginalized people to make sense of, and claim agency within, their socio-historical conditions" (2014: 265).

Harsh anti-immigrant discourse has marked undocumented people as intruders, invaders, and aliens. The repetitive use of such terminology in the media and political discourse often leads to its internalization by young people. Navigating illegality thus is essential in acknowledging the daily practices of undocumented youth and is often helpful in the process of becoming part of political action and overcoming the fear and shame of coming out (Enriquez and Saguy 2016; Negrón-Gonzales 2013).

Formal legal status, or lack thereof, powerfully shapes the identity of undocumented migrants, particularly youth and children. A lack of status disengages their participation in both community and political activities, at least to a certain extent. Their deportability relegates many families to the margins of their communities. This fear impacts both on the "practice of cultural citizenship and their notions of belonging" (Gonzales et al. 2015: 331). However, in the case of the undocumented migrant youth in the United States, the visibility of their plight (Gonzales et al. 2015: 332) and the use of the cultural schema of coming out has allowed them to allay their fears by revealing their immigration status while promoting their political mobilization (Enriquez and Saguy 2016). Counter-storytelling, providing and expressing first-person narratives, enables young people to construct an identity that is true to them (Gonzales et al. 2015).

Constructing frames of deservingness

Contemporary citizenship is a dynamic process shaped by both macro-level structures (global processes and immigration processes) and micro-level practices (the everyday decisions and actions of migrants) (Gonzales and Sigona 2017: 6). As we have discussed in previous chapters, in recent years there has been a considerable multiplication of the different possible immigration statuses that states afford to migrants. In addition, migrants move from one category to another and, in any case, access to broader or fewer rights depends on their immigration status. While some migrants have temporary immigration status, they may live alongside those who have no form of legal status and others who have full forms of legal status. This legal spectrum is influenced by how immigration policies are made and whether some of the grayer areas are kept in place or eliminated, as is the case with TPS and DACA.

In restrictive and hostile immigration environments, it is not only undocumented migrants who experience increasing hardship. This can spread to legalized migrants who see their status challenged by changes in immigration law and policy. In their comparative study of migrant youth mobilization in the Netherlands and the United States, Walter Nicholls and colleagues (Nicholls, Maussen, and de Mesquita 2016) explore the similarities in the discursive practices that migrant youth in both countries use in their campaigns. The authors argue that the resulting similarities emerge from the convergence of the citizenship regimes in these two national contexts. In both cases, the campaigns tie ideas of deservingness of these migrant youths to specific cultural attributes. These attributes are features that might make immigrant youths indistinguishable from their citizen peers. For example, in the case of the Netherlands, by removing the headscarf, a young Muslim immigrant woman may claim her cultural adaptation, which then makes her seem more "worthy." To the wider society, then, the moral dilemma that such campaigns cultivate concerns the putative deportation of people who speak, look, and behave "normal," "like us." According to the authors, restrictive immigration policy environments in these countries generate *niche-openings* for complying immigrants (Nicholls, Maussen, and de Mesquita 2016: 1594) and give rise to politics of deservingness.

Similarly, the DREAMer movement in the United States has drawn heavily on the deservingness of this population to make claims about their rights to remain in the country. As Nicholls (2013) captures, in targeting a public largely hostile to undocumented migrants, the organizers of the movement crafted an image of these immigrant youths as exceptional academic achievers, model "citizens" of their communities, and deserving of greater rights and a pathway to permanent residency (Nicholls and Fiorito 2015: 2).

Advocates for these young people argue that they were brought to the United States through "no fault of their own" and thus are worthy of greater opportunities based on their innocence. The campaign, however, had the counter-effect of reinforcing the cleavage between deserving and undeserving migrants. For instance, the strategy unwillingly blamed their parents for bringing them into the United States, placing the onus on their parents for the "original sin."

However, undocumented young people became troubled by the ways in which the political debates were contrasting their deservingness with their parents' guilt and attempted to shift the discourse. But by then the saying "through no fault of their own" had been used by various advocates, picked up by the mainstream media, and spread everywhere. As some of the young migrants and the more critical activists felt increasingly uncomfortable with this strategy, other more radical tactics were employed in many cities: occupations of government offices, hunger strikes, marches, and other forms of direct action. This insurgent stream in the DREAMer movement explicitly aimed at blurring the boundaries between these young people and the rest of the migrant population and at stressing broader identities cutting across movements such as labor and LGBTQ rights.

In 2013, University of California Berkeley graduate Ju Hong, an undocumented Korean immigrant from the Bay Area, stood up during a speech by President Obama at the Betty Ong Center in San Francisco. Hong, who migrated to the United States at the age of eleven, shouted over the president, "I need your help. My family will be separated on Thanksgiving. Please use your executive [power]. You have power to stop deportations."

As we can see from the earlier efforts of DREAMers and their advocates, the politics of deservingness can reproduce negative stereotypes of undocumented migrants and further consolidate the differences between them and more "worthy" subjects. But, most importantly, these politics of deservingness produce a certain kind of deserving migrant, a subject that is docile, constantly proving his or her "worthiness" by working hard, behaving, and staying out of trouble. Drawing on a longitudinal study with forty young people of Mexican descent in San Francisco, Irene Bloemraad and her colleagues (Bloemraad, Sarabia, and Fillingim 2016) find that, while the parents' undocumented status did not seem to have a negative impact on their children's civic engagement and political participation, young people felt tremendous pressure to stay out of trouble. These young people face a dual pressure which requires them to protect their families by not standing out while also making formal claims on their families' behalf in the political system (2016: 1546–7).

Political socialization within the family

Similarly, the idea of political socialization – how individuals acquire their political orientations through knowledge and attitudes – provides interesting insights into the ways in which engagement of migrant youth in politics may be initiated. Sociologist Veronica Terriquez finds that barriers related to undocumented status, in effect, reduce migrant parents' political engagement and suppress the civic and political participation of their US-raised children, supporting a top-down model of political socialization, where the parent is the source of a child's political awareness (Terriquez and Kwon 2015: 425). However, she finds that some young people leverage experience in youth-organizing efforts to encourage their parents' participation and invoking political socialization of the family. As such, this new knowledge "trickles up" to their parents.

Adult migrants can be disconnected from political participation due to language barriers, low educational attainment, or lack of US citizenship (Terriquez and Kwon 2015). However, evidence shows that witnessing their children's increased interest in politics and activism may stimulate a parent's interest (Terriquez and Kwon 2015; Wong and Tseng 2008). Youth who obtain political knowledge are very likely to share this with their families and, in this way, serve as "politicizing agents" (Terriquez and Kwon 2015).

Wong and Tseng (2008) argue that the children of migrants residing in the United States are more likely to be exposed to political structures and information, and thus they are likely to be a critical source of political information. They also suggest that political socialization could be bidirectional. That is, both young people and their parents may transfer political information and initiate a dialogue of ideas and opinions. Some research, albeit not extensive yet, has been done on this analysis of bidirectional political socialization focusing on migrant families. Proficiency in English is often a barrier to migrant parents' participation in and engagement with politics. However, due to cultural and linguistic proficiencies, second-generation migrant children play a critical role in translating for their families and overcoming this barrier. This is how second-generation youth become exposed to political institutions and local and state agencies. As Negrón-Gonzales found in her research, "children of immigrants serve not only as translators in the literal sense but are also frequently called upon to broker their parents' daily interaction with unfamiliar, social, political, and institutional terrain the accompanies daily life" (2014: 266). Children's awareness of their parents' legal status is also a factor in deconstructing political socialization within migrant families.

Irene Bloemraad and her colleagues (Bloemraad, Sarabia, and Fillingim 2016) offer insights into the question of whether parents' immigration status has any effect on the political and civic engagement of their second-generation children. The authors found that there was no evidence of "apathy of disengagement" (2016: 1534). On the contrary, the majority of the youth they interviewed had joined the spring 2006 migrant-rights marches across the United States. They also participated in community organizations and volunteer opportunities. However, some families displayed caution and advised their children to "stay out of trouble" so as not to draw attention to themselves and the family (Bloemraad et al. 2016: 1534).

US citizens who are members of mixed-status households are more likely to be personally involved with the struggles associated with having undocumented parents and are also more likely to have their political opinions and participation shaped by this experience. A recent study conducted by Amuedo-Dorantes and Lopez (2017) examines whether the increasingly restrictive immigration policies and intensified enforcement practices have effectively weakened the political engagement of US citizens living in mixed-status families. The authors found that individuals with foreign-born parents were more likely to show their discontent for the way that undocumented migrants are being treated and thus were more likely to participate in protests, actions, and marches.

Finally, as Roberto G. Gonzales and his colleagues argue, the public education system plays a crucial role in students' political and civic development. Schools function as "integrators, as constructors of citizenship, and as facilitators of public and community engagement" (Gonzales et al. 2015: 318). This means that schools shape the civic and political experiences of undocumented students, as well as the opportunities available to them. Integration into mainstream US culture also happens through the educational system: "inclusion in schools function[s] to give [students] access to a state-sanctioned space of inclusion in the polity as well as the tools through which to bolster their claims to broader frames of membership, extending beyond the school and into community and society more generally" (Gonzales et al. 2015: 326). The school experience is the main driver shaping their narratives and assisting them to become politically engaged and active. Yet the limitations that accompany students' undocumented status and the social stratification associated with underfunded schools present barriers to migrant students' civic education. It may prevent them from becoming fully engaged and dampen their pursuits of membership (Gonzales et al. 2015).

Conclusion

On February 1, 2019, after a hundred days in a private detention facility, undocumented migrant and activist Eduardo Samaniego was deported to Mexico (Christensen 2019). In October of 2018, Eduardo was arrested in his home state of Georgia after forgetting his wallet and failing to pay a cab fare, a misdemeanor charge. Because he was undocumented, the situation took a dramatic turn. He was taken to jail and then transferred to the custody of ICE. While in detention he spent several days in solitary confinement, causing him mental health strain.

Eduardo, who migrated to the United States at the age of sixteen, graduated from high school with honors and was president of Junior Achievement in Georgia, president of the Hispanic Honor Society, and the only National Society of High School Scholars gold-medal winner in his graduating class. After being denied admission to universities in his home state of Georgia, due to the policy excluding undocumented migrants from Georgia's most competitive public universities, Hampshire College, a private liberal arts institution in Amherst, Massachusetts, offered him a four-year scholarship.

Eduardo was known as an immigrant-rights organizer in Massachusetts and Georgia, organizing for the Pioneer Valley Workers Center and Freedom University. Due to the age at which he migrated, he was not eligible for DACA. It is unclear whether Eduardo's detention and subsequent deportation were driven by his activism. But Eduardo's story sheds light on some of the debates we have wrestled with in this book. His story is one of membership and belonging and of exclusion and expulsion. Like many undocumented migrants, his story is firmly located in the places where he grew up, where he was denied education, where he went to college, and where he was ultimately arrested and detained. Many of his friends would likely attest to his success as an immigrant rights organizer. Following his arrest and detainment, Eduardo had an impressive network of supporters. But ultimately, he faced a punitive enforcement system that negatively impacted his health and expelled him from the country.

This book has debated the conflicting and contradictory experiences faced by migrants in their attempts to establish a home and to assert belonging in the face of increasingly hostile immigration controls. In her work on undocumented migrants, Susan Bibler Coutin argues that the contradictions between undocumented migrants' physical and social presence and their official designation as "illegal" generate "spaces of nonexistence" (Coutin 2000: 27–47). In a similar consideration of the dilemmas that this book raises, Coutin declares:

> I cannot celebrate the space of nonexistence. Even if this space is in some ways subversive, even if its boundaries are permeable, and even if it is sometimes irrelevant to individuals' everyday lives, nonexistence can be deadly. Legal nonexistence can mean being detained and deported, perhaps to life-threatening conditions. It can mean working for low wages in a sweatshop, or being unemployed. It can mean the denial of medical care, food, social service, education, and public housing. And it can mean an erasure of rights and personhood such

that violence becomes not only legitimate but even required. (Coutin 2003: 193)

This particular chapter in Eduardo Samaniego's story upholds the enduring power of the nation-state and demonstrates the "master status" quality of illegality. Yet, on the eve of his deportation, dozens of supporters – citizens and noncitizens alike – gathered at the detention center in South Carolina to say goodbye. Perhaps Eduardo will not return to the United States. But, as this immigrant rights movement has shown, there are many to persevere in his absence.

Notes

Chapter 1 Who Are Undocumented Migrants?

1 Extracts from interview collected for the ESRC-funded Becoming Adult project (www.becomingadult.net) led by the University College London in partnership with the University of Birmingham and the University of Oxford.
2 A growing number of countries, including the United States and Australia, have entered into agreements with neighboring countries for the removal of unauthorized border crossers, along with offshore immigration processing and detention facilities.
3 Unravelling the Mediterranean Migration Crisis (MEDMIG) is a study led by Coventry University in partnership with the Universities of Birmingham and Oxford on the drivers of the so-called "refugee crisis" (2015–16). The study was funded by the Economic and Social Research Council and the Department for International Development (DFiD).
4 For a discussion of how different terms and labels intervened in framing the so-called "refugee crisis" in the Mediterranean region, see Crawley and Skleparis 2018; Sigona 2018.
5 Bart Jensan, John Fritze, and Alan Gomez. "White House approves military to use lethal force at southern border," *USA Today*, November 21, 2018.
6 We recognize that changes to immigration policies that might move migrants across categories can make it difficult to generate a profile of undocumented immigrants. We are also cognizant of the political processes that drive processes of labeling and categorization. As such, we acknowledge that the ways in which migrants are enumerated reflect decisions that are deeply political.
7 Although this may seem an unlikely scenario, it is in the realm of legal possibility, as shown by recent research on the impact of the Brexit referendum on EU nationals in the United Kingdom (Yeo 2018).
8 In the United States, joining DACA beneficiaries are more than 300,000

immigrants from El Salvador, Haiti, Honduras, Nepal, Nicaragua, and Sudan who have lived in the United States under a Temporary Protected Status (TPS) designation.
9 Defined as people residing in a country other than their country of birth.
10 Global Migration Data Analysis Centre, 2018, Global Migration Indicators 2018: Insights from the Global Migration Data Portal. http://publications.iom.int/system/files/pdf/global_migration_indicators_2018.pdf
11 Jens Manuel Krogstad, Jeffrey S. Passel, and D'Vera Cohn, "Five Facts about Illegal Immigration in the US," Pew Research Center, November 28, 2018. http://www.pewresearch.org/fact-tank/2018/11/28/5-facts-about-illegal-immigration-in-the-u-s/
12 Ibid.
13 Marc R. Rosenblum and Ariel G. Ruiz Soto, "An Analysis of Unauthorized Immigrants in the United States by Country and Region of Birth," Migration Policy Institute, August 2015. https://www.migrationpolicy.org/research/analysis-unauthorized-immigrants-united-states-country-and-region-birth
14 Clandestino Project Final Report, November 23, 2009. http://clandestino.eliamep.gr/wp-content/uploads/2010/03/clandestino-final-report_-november-2009.pdf
15 Jo Woodbridge, "Sizing the Unauthorized (Illegal) Migrant Population in the United Kingdom in 2001," Home Office Online Report, 2005, http://clandestino.eliamep.gr/wp-content/uploads/2010/03/clandestino-final-report_-november-2009.pdf. See also Vollmer 2011.
16 J. Simpson and E. S. Weiner (1989), *Oxford English Dictionary* online. Oxford: Clarendon Press.
17 This figure includes children born in Italy to immigrant parents but does not include those who have acquired Italian nationality.

Chapter 2 Theorizing the Lived Experience of Migrant Illegality

1 In his classic definition of citizenship – "full membership of the community, with all its rights and responsibilities" – British sociologist T. H. Marshall does not mention the state. See Marshall 1950.
2 As Leo Chavez (2008: 22) points out, dominant discourse about undocumented Mexicans has emerged from "a history of ideas, laws, narratives, myths, and knowledge production in social sciences, sciences, the media, and the arts."

Chapter 3 Geographies of Undocumented Migration

1 https://www.theguardian.com/politics/2009/mar/10/boris-immigration-london
2 https://www.ft.com/content/18bf3b42-6eda-11e2-8189-00144feab49a

3 Established in 1996, the Schengen Area is an area comprising 26 European states, of which 22 EU member states have officially abolished passports and all other types of border control at their mutual borders. The area mostly functions as a single jurisdiction for international travel purposes, with a common visa policy.
4 The New York City Identity Card provides access to a number of services including "a free one-year membership of many of the city's leading museums, zoos, concert halls, and botanical gardens," the ability to open a bank account with participating banks and credit unions, but more importantly it serves as a recognized form of identification for interacting with the New York Police Department. https://www1.nyc.gov/site/idnyc/index.page
5 http://www1.nyc.gov/site/immigrants/about/message-from-commissioner.page
6 https://www.nbcnews.com/politics/immigration/six-states-nyc-sue-trump-admin-over-requiring-sanctuary-cities-n892596https://www.washingtonpost.com/news/post-nation/wp/2018/08/01/trumps-order-threatening-to-withhold-funding-from-sanctuary-cities-is-unconstitutional-court-rules/?utm_term=.6bccbfb21112
7 https://www.theguardian.com/world/2017/feb/18/protesters-in-barcelona-urge-spain-to-take-in-more-refugees
8 https://www.bbc.co.uk/news/world-europe-35854413
9 http://solidaritycities.eu/about

Chapter 4 Immigration Enforcement, Detention, and Deportation

1 Maria Sacchetti, "ICE raids meatpacking plant in rural Tennessee; 97 immigrants arrested," *Washington Post*, April 6, 2018.
2 Interviews collected as part of the project Undocumented Migrant Children and Families in Britain (Sigona and Hughes 2012).
3 In 2012, after a two-year court battle that struck down some of the other provisions, a federal court judge approved implementation of the "Show me your papers" provision of S. B. 1070. See Santos (2012).
4 The National Immigration Law Center compiles a running list of anti-immigrant legislation and its status in state legislatures and in the courts.
5 The British Nationality Act of 1948 imparted the status of citizenship of the United Kingdom and colonies to all British subjects connected with the United Kingdom or British colonies.

Chapter 5 Undocumented Status and Social Mobility

1 Between 2013 and 2017, the California legislature passed seven laws designed to protect undocumented immigrants in the state from retaliation and discrimination related to their immigration status: AB 263 (2013), SB

Notes to pp. 111–133

666 (2013), AB 524 (2013), AB 2751 (2014), AB 622 (2015), SB 1001 (2016), and AB 450 (2017).
2 The principle, "best interests of the child," derives from Article 3 of the United Nations Convention on the Rights of the Child, which states that "in all actions concerning children, whether undertaken by public or private social welfare institutions, courts of law, administrative authorities or legislative bodies, the best interests of the child shall be a primary consideration."
3 Quoted from Sigona and Hughes 2012.

Chapter 6 Families and Children

1 As an administrative memorandum that shifts bureaucratic practice in US Customs and Border Protection, US Customs and Immigration Services (USCIS), and US Immigration and Customs Enforcement, DACA has limited inclusionary power; it does not offer a pathway to citizenship or other forms of legal status. It also does not lift exclusions from federal financial aid. Its temporary and partial protections create a two-year (renewable) prosecutorial discretion with regard to deportation. In addition, its beneficiaries can also receive work authorization and apply for a social security card.
2 Texas Attorney General Ken Paxon, along with nine other state attorneys general, sent a letter to US AG Sessions threatening to sue unless the Trump administration ended the program by September 5, 2017.
3 Tal Kopan, "Trump ends DACA but gives Congress window to save it," *CNN*, September 5, 2017.
4 Interview conducted by the National UnDACAmented Research Project, Roberto G. Gonzales, PI.
5 http://eur-lex.europa.eu/legal-content/EN/TXT/PDF/?uri=CELEX:32003L0086&from=EN
6 http://picum.org/wp-content/uploads/2017/11/Children-Testimonies_EN.pdf

References

Abrego, L. (2006). "I Can't Go to College because I Don't have Papers": Incorporation Patterns of Latino Undocumented Youth. *Latino Studies* 4(3): 212–31.

Abrego, L. (2008). Legitimacy, Social Identity, and the Mobilization of Law: The Effects of Assembly Bill 540 on Undocumented Students in California. *Law & Social Inquiry* 33(3): 709–34. https://doi.org/10.1111/j.1747-4469.2008.00119.x

Abrego, L. (2009). Economic Well-being in Salvadoran Transnational Families: How Gender Affects Remittance Practices. *Journal of Marriage and the Family* 71(4): 1070–85.

Abrego, L. J. (2011). Legal Consciousness of Undocumented Latinos: Fear and Stigma as Barriers to Claims-Making for First- and 1.5-Generation Immigrants. *Law & Society Review* 45(2): 337–69.

Abrego, L. J. (2014). *Sacrificing Families: Navigating Laws, Labor, and Love across Borders*. Redwood City, CA: Stanford University Press.

Abrego, L. J. (2016). Illegality as a Source of Solidarity and Tension in Latino Families. *The Journal of Latino / Latin American Studies; Omaha* 8(1): 5–21.

Agamben, G. (1998). *Homo Sacer: Sovereign Power and Bare Life*. Redwood City, CA: Stanford University Press.

Aleinikoff, T. A. (1997). *Citizen and Membership: A Policy Perspective*. Washington, DC: Carnegie Endowment for International Peace.

Allard, S. W. (2009). *Out of Reach: Place, Poverty, and the New American Welfare State*. New Haven, CT: Yale University Press.

Allard, S. and Roth, B. (2010). Strained Suburbs: The Social Service Challenges of Rising Suburban Poverty, Policy File. Brookings Metropolitan Policy Program.

Allsopp, J. and Chase, E. (2017). Best Interests, Durable Solutions and Belonging: Policy Discourses Shaping the Futures of Unaccompanied Migrant and Refugee Minors Coming of Age in Europe. *Journal of Ethnic and Migration Studies* 45(2): 293–311. https://doi.org/10.1080/1369183X.2017.1404265

References

Allsopp, J., Chase, E., and Mitchell, M. (2015). The Tactics of Time and Status: Young People's Experiences of Building Futures While Subject to Immigration Control in Britain. *Journal of Refugee Studies* 28(2): 163–82. https://doi.org/10.1093/jrs/feu031

Alonzo, A., Macleod-Ball, K., Chen, G., and Kim, S. (2011) Enforcement Off Target: Minor Offenses with Major Consequences. American Immigration Lawyers Association. https://www.aila.org/File/Related/11081609.pdf

Amoore, L., Marmura, S., and Salter, M. B. (2008). Smart Borders and Mobilities: Spaces, Zones, Enclosures. *Surveillance & Society* 5(2).

Amuedo-Dorantes, C., and Lopez, M. (2017). Interior Immigration Enforcement and Political Participation of US Citizens in Mixed-Status Households. *Demography* 54(6): 2223–47.

Anderson, B. (2006). *Imagined Communities: Reflections on the Origin and Spread of Nationalism.* London and New York: Verso Books.

Anderson, B. and Ruhs, M. (2010). Researching Illegality and Labour Migration. *Population, Space and Place* 16(3): 175–9. doi: 10.1002/psp.594

Andersson, R. (2014). *Illegality, Inc.: Clandestine Migration and the Business of Bordering Europe.* Berkeley, CA: University of California Press.

Andreas, P. (2009). *Border Games: Policing the US–Mexico Divide*, 2nd edn. Ithaca: Cornell University Press.

Ansley, F. (2010). Constructing Citizenship without a Licence: The Struggle of Undocumented Immigrants in the USA for Livelihoods and Recognition. *Studies in Social Justice* 4(2): 165–78.

Arce, J. (2016). *My (Underground) American Dream: My True Story as an Undocumented Immigrant Who Became a Wall Street Executive.* New York: Hachette Nashville.

Arnett, J. J. (2000). Emerging Adulthood: A Theory of Development from the Late Teens through the Twenties. *American Psychologist* 55(5): 469–80. https://doi.org/10.1037/0003-066X.55.5.469

Artico, C. I. (2003). *Latino Families Broken by Immigration: The Adolescent's Perceptions.* New York: LFB Scholarly Publishing.

Badiou, A. (2009). *Theory of the Subject.* London: A&C Black.

Balibar, E. (1996). Is European Citizenship Possible? *Public Culture* 8(2): 355–76.

Balibar, E. (1988). *Race, Nation, Class: Ambiguous Identities.* London and New York: Verso.

Barrett, J., and Sigona, N. (2014). The Citizen and the Other: New Directions in Research on the Migration and Citizenship Nexus. *Migration Studies* 2(2): 286–94. https://doi.org/10.1093/migration/mnu039

Bauder, H. and Gonzalez, D. (2018). Municipal Responses to "Illegality": Urban Sanctuary across National Contexts. *Social Inclusion* 6(1): 124–34.

Bean, F. D. and Stevens, G. (2003). *America's Newcomers and the Dynamics of Diversity.* New York: Russell Sage Foundation.

Bean, F. D., Leach, M. A., Brown, S. K., Bachmeier, J. D., and Hipp, J. R.

References

(2011). The Educational Legacy of Unauthorized Migration: Comparisons across US-Immigrant Groups in How Parents' Status Affects Their Offspring. *International Migration Review* 45(2): 348–85. https://doi.org/10.1111/j.1747-7379.2011.00851.x

Bean, F. D., Telles, E. E. and Lowell, L. B. (1987). Undocumented Migration to the United States: Perceptions and Evidence. *Population and Development Review* 13(4): 671–90.

Becker, H. S. (1963). *Outsiders: Studies in the Sociology of Deviance*. London: Free Press of Glencoe. http://nrs.harvard.edu/urn-3:hul.ebookbatch.ASP_batch:ASPS100020155soth

Benton, L. (1994). Beyond Legal Pluralism: Towards a New Approach to Law in the Informal Sector. *Social & Legal Studies* 3(2): 223–42.

Berlin, G., Furstenberg, F., and Waters, M. C., (2010). The Transition to Adulthood: Introducing the Issue. http://nrs.harvard.edu/urn-3:HUL.InstRepos:33439193

Bermúdez, A. and Escrivá, Á. (2016). La participación política de los inmigrantes en España: Elecciones, representación y otros espacios. *Anuario CIDOB De La Inmigración* 2015-2016, pp. 296–317.

Bickham, J. and Nelson, L. (2016) Producing 'Quality of Life' in the 'Nuevo South': The Spatial Dynamics of Latinos' Social Reproduction in Southern Amenity Destinations *City and Society* 28(2): 129–51.

Bigo, D. (2001). The Möbius Ribbon of Internal and External Security(ies), in M. Albert, D. Jacobson, and Y. Lapid, *Identities, Borders, Orders: Rethinking International Relations Theory*. Minneapolis, MN: University of Minnesota Press, pp. 91–116.

Bloch, A. and McKay, S. (2016). *Living on the Margins: Undocumented Migrants in a Global City*. Bristol: Policy Press.

Bloch, A. and Schuster, L. (2005). At the Extremes of Exclusion: Deportation, Detention and Dispersal. *Ethnic and Racial Studies* 28(3): 491–512. https://doi.org/10.1080/0141987042000337858

Bloch, A., Sigona, N., and Zetter, R. (2014) *Sans Papiers: The Social and Economic Lives of Undocumented Migrants*. London and New York: Pluto Press.

Bloemraad, I., Sarabia, H., and Fillingim, A. E. (2016). "Staying Out of Trouble" and Doing What Is "Right": Citizenship Acts, Citizenship Ideals, and the Effects of Legal Status on Second-Generation Youth. *American Behavioral Scientist* 60(13): 1534–52.

Boehm, D. A. (2012). *Intimate Migrations: Gender, Family, and Illegality among Transnational Mexicans*. New York: New York University Press. http://nrs.harvard.edu/urn-3:hul.ebookbatch.PMUSE_batch:muse9780814789858

Boehm, D. A. (2016). *Returned: Going and Coming in an Age of Deportation*. Berkeley, CA: University of California Press.

Boehm, D. A. (2017). Separated Families: Barriers to Family Reunification after

References

Deportation. *Journal on Migration and Human Security* 5(2). http://search.proquest.com/docview/1931569186/abstract/162CF393AEF84974PQ/1

Bonefeld, W. (1994). *The Recomposition of the British State during the 1980s.* Brookfield, VT: Dartmouth.

Bonefeld, W. (1995). Capital as Subject and the Existence of Labour, in W. Bonefeld, R. Gunn, J. Holloway, and K. Psychopedis (eds), *Emancipating Marx: Open Marxism 3*, East Haven, CT: Pluto.

Borjas, G. J. (1987). Self-Selection and the Earnings of Immigrants. *The American Economic Review* 77(4): 531–53.

Bosniak, L. (2000) Citizenship Denationalized (The State of Citizenship Symposium). *Indiana Journal of Global Legal Studies* 7(2), Article 2. https://www.repository.law.indiana.edu/ijgls/vol7/iss2/2

Bosniak, L. (2006). *The Citizen and the Alien: Dilemmas of Contemporary Membership*. Princeton: Princeton University Press.

Boswell, C. (2007). Theorizing Migration Policy: Is There a Third Way? *International Migration Review* 41: 75–100. doi: 10.1111/j.1747-7379.2007.00057.x

Bourdieu, P. and Wacquant, L. J. (1992). *An Invitation to Reflexive Sociology*. Chicago, IL: University of Chicago Press.

Bowles, S. and Gintis, H. (1976). *Schooling in Capitalist America*. New York: Basic Books.

Brigden, N. (2018). *The Migrant Passage: Clandestine Journeys from Central America*. Ithaca, NY: Cornell University Press.

Burawoy, M. (1976) The Functions and Reproduction of Migrant Labor: Comparative Material from Southern Africa and the United States. *American Journal of Sociology* 81(5): 1050–87.

Buriel, R. and De Ment, T. (1997) Immigration and Sociocultural Change in Mexican, Chinese, and Vietnamese American Families, in A. Booth, A. C. Crouter, and N. S. Landale (eds), *Immigration and the Family: Research and Policy on US Immigrants*. Hillsdale, NJ: Lawrence Erlbaum Associates, pp. 165–200.

Bustamante, J. (1976) Structural and Ideological Conditions of the Mexican Undocumented Immigration to the United States. *American Behavioral Scientist* 19(3): 364–76.

Calavita, K. (1998). Immigration, Law, and Marginalization in a Global Economy: Notes from Spain. *Law & Society Review* 32(3): 529–66.

Calavita, K, (2005). *Immigrants at the Margins: Law, Race, and Exclusion in Southern Europe*. New York: Cambridge University Press.

Canizales, S. L. (2015). American Individualism and the Social Incorporation of Unaccompanied Guatemalan Maya Young Adults in Los Angeles. *Ethnic and Racial Studies* 38(10): 1831–47. https://doi.org/10.1080/01419870.2015.1021263

Canizales, S. L. (2018). Support and Setback: How Religion and Religious

References

Organisations Shape the Incorporation of Unaccompanied Indigenous Youth. *Journal of Ethnic and Migration Studies* 1–18. https://doi.org/10.1080/1369183X.2018.1429899

Capps, R., Fix, M., Murray, J. et al. (2005). The New Demography of America's Schools: Immigration and the No Child Left Behind Act. Urban Institute (NJ1).

Carliner, G. (1980). Wages, Earnings and Hours of First, Second, and Third Generation American Males. *Economic Inquiry* 18(1): 87–102. https://doi.org/10.1111/j.1465-7295.1980.tb00561.x

Carling, J. (2002) Migration in the Age of Involuntary Immobility: Theoretical reflections and Cape Verdean Experiences, *Journal of Ethnic and Migration Studies* 28(1):5-42. DOI: 10.1080/13691830120103912

Carroll, M. S. (1988). A Tale of Two Rrivers: Comparing NPS-local Interactions in Two Areas. *Society & Natural Resources* 1(1): 317–33. https://doi.org/10.1080/08941928809380663

Carter, D. M. (1997). *States of Grace: Senegalese in Italy and the New European Immigration*. Minneapolis, MN: University of Minnesota Press.

Castles, S. (2004). Why Migration Policies Fail. *Ethnic & Racial Studies* 27(2): 205–27. https://doi.org/10.1080/0141987042000177306

Castles, S. and Davidson, A. (2000). *Citizenship and Migration: Globalization and the Politics of Belonging*. Basingstoke: Macmillan.

Castro-Salazar, R., and Bagley, C. (2010). "Ni de aquí ni from there." Navigating between Contexts: Counter-narratives of Undocumented Mexican Students in the United States. *Race Ethnicity and Education* 13(1): 23–40.

Cebulko, K. and Silver, A. (2016). Navigating DACA in Hospitable and Hostile States: State Responses and Access to Membership in the Wake of Deferred Action for Childhood Arrivals. *American Behavioral Scientist* 60(13): 1553–74. https://doi.org/10.1177/0002764216664942

Chacón, J. M. (2012). Overcriminalizing Immigration. *Journal of Criminal Law and Criminology* 102(3): 613–52.

Chacón, J. (2014). Immigration Detention: No Turning Back? *South Atlantic Quarterly* 113(3): 621–8.

Charles, M. (1982). The Yellowstone Ranger: The Social Control and Socialization of Federal Law Enforcement Officers. *Human Organization* 41(3): 216–26. https://doi.org/10.17730/humo.41.3.p187n30002335677

Chase, E. (2017). Health and Wellbeing. Becoming Adult Research Brief No. 5, London: UCL. www.becomingadult.net

Chauvin, S. and Garcés-Mascareñas, B. (2012). Beyond Informal Citizenship: The New Moral Economy of Migrant Illegality. *International Political Sociology* 6(3): 241–59.

Chauvin, S. and Garcés-Mascareñas, B. (2014). Becoming Less Illegal: Deservingness Frames and Undocumented Migrant Incorporation. *Sociology Compass* 8(4): 422–32. https://doi.org/10.1111/soc4.12145

References

Chavez, L. R. (1988). Settlers and Sojourners – The Case of Mexicans in the United States. *Human Organization* 47(2): 95–108.

Chavez, L. R. (1991). *Shadowed Lives: Undocumented Immigrants in American Society*. Cengage Learning.

Chavez, L. (2007). The Condition of Illegality. *International Migration* 45(3): 192–6.

Chavez, L. R. (2008). *The Latino Threat: Constructing Immigrants, Citizens, and the Nation*. Redwood City, CA: Stanford University Press.

Chen, A. and Gill, J. (2015). Unaccompanied Children and the US Immigration System: Challenges and Reforms. *Journal of International Affairs* 68(2): 115–36.

Chiswick, B. R. (1991). Speaking, Reading, and Earnings among Low-skilled Immigrants. *Journal of Labor Economics* 9(2): 149–70.

Christensen, D. (2019). Pioneer Valley Activist Eduardo Samaniego Deported to Mexico. *Daily Hampshire Gazette*, February 1, 2019. https://www.gazettenet.com/Local-activist-Eduardo-Samaniego-deported-to-Mexico-23195534

Coleman, J. S. (1988). Social Capital in the Creation of Human Capital. *American Journal of Sociology* 94: S95–S120.

Coleman, M. (2007). Immigration Geopolitics beyond the Mexico–US Border. *Antipode*, 39(1): 54–76.

Coleman, M. (2012). The "Local" Migration State: The Site-Specific Devolution of Immigration Enforcement in the US South. *Law & Policy* 34(2): 159–90. https://doi.org/10.1111/j.1467-9930.2011.00358.x

Coleman, M. and Stuesse, A. (2016). The Disappearing State and the Quasi-Event of Immigration Control. *Antipode* 48(3): 524–43.

Coll, K. (2010). *Remaking Citizenship: Latina Immigrants and New American Politics*. Redwood City, CA: Stanford University Press.

Contreras, F. (2009). Sin Papeles y Rompiendo Barreras: Latino Students and the Challenges of Persisting in College. *Harvard Educational Review* 79(4): 610–32.

Cooper, K. (2018). Patient Information Shared with Immigration Officials. British Medical Association, January 26, 2018. https://www.bma.org.uk/news/2018/january/patient-information-shared-with-immigration-officials

CorporateWatch (2016) Snitches, Stings, and Leaks: How "Immigration Enforcement" Works. London. https://corporatewatch.org/snitches-stings-leaks-how-immigration-enforcement-works-2/

Côté, J. E. and Allahar, A. (1996) *Generation on Hold: Coming of Age in the Late Twentieth Century*. New York and London: New York University Press.

Coutin, S. (1993). *The Culture of Protest: Religious Activism and the US Sanctuary Movement* (Conflict and Social Change series). Boulder, CO: Westview Press.

Coutin, S. (1999). Citizenship and Clandestiny among Salvadoran Immigrants. *PoLAR: Political and Legal Anthropology Review* 22(2): 53–63.

Coutin, S. (2000). *Legalizing Moves: Salvadoran Immigrants' Struggle for US Residency*. Ann Arbor, MI: University of Michigan Press.

References

Coutin, S. B. (2003). Illegality, Borderlands, and the Space of Nonexistence, in R. W. Perry and B. Maurer (eds), *Globalization under Construction: Governmentality, Law, and Identity*. Minneapolis: University of Minnesota Press, pp. 171–202.

Coutin, S. (2011). The Rights of Noncitizens in the United States. *Annual Review of Law and Social Science* 7(1): 289–308. https://doi.org/10.1146/annurev-lawsocsci-102510-105525

Coutin, S. (2013). In the Breach: Citizenship and its Approximations. *Indiana Journal of Global Legal Studies* 20(1): 109–40.

Crawley, H. and Skleparis, D. (2018). Refugees, Migrants, Neither, Both: Categorical Fetishism and the Politics of Bounding in Europe's "Migration Crisis." *Journal of Ethnic and Migration Studies* 44(1): 48–64.

Crawley, H., Duvell, F., Jones, K., McMahon, S., and Sigona, N. (2017). *Unravelling Europe's "Migration Crisis."* Bristol: Policy Press.

Cvajner, M. and Sciortino, G. (2010). Theorizing Irregular Migration: The Control of Spatial Mobility in Differentiated Societies. *European Journal of Social Theory* 13(3): 389–404. https://doi.org/10.1177/1368431010371764

Dalakoglou, D. (2013). "From the Bottom of the Aegean Sea" to Golden Dawn: Security, Xenophobia, and the Politics of Hate in Greece. *Studies in Ethnicity and Nationalism* 13(3): 514–22.

Dalla, R. L., Ellis, A., and Cramer, S. C. (2005). Immigration and Rural America. *Community, Work & Family* 8(2): 163–85. https://doi.org/10.1080/13668800500049639

Darling, J. (2009). Becoming Bare Life: Asylum, Hospitality, and the Politics of Encampment. *Environment and Planning D: Society and Space* 27(4): 649–65. https://doi.org/10.1068/d10307

De Genova, N. P. (2002). Migrant "Illegality" and Deportability in Everyday Life. *Annual Review of Anthropology* 31: 419–47.

De Genova, N. (2005). *Working the Boundaries: Race, Space, and "Illegality" in Mexican Chicago*. Durham, NC: Duke University Press.

De Genova, N. (2009). Conflicts of Mobility, and the Mobility of Conflict: Rightlessness, Presence, Subjectivity, Freedom. *Subjectivity* 29(1): 445–66.

De Genova, N. and Peutz, N. (2010). *The Deportation Regime: Sovereignty, Space, and the Freedom of Movement*. Durham, NC: Duke University Press.

De Graauw, E. (2016). *Making Immigrant Rights Real: Nonprofits and the Politics of Integration in San Francisco*. Ithaca, New York and London: Cornell University Press.

DeLanda, M. (2016). *Assemblage Theory*. Edinburgh: Edinburgh University Press.

De León, J. (2015). *The Land of Open Graves: Living and Dying on the Migrant Trail*. Berkeley, CA: University of California Press.

Deleuze, G. and Guattari, F. (1988). *A Thousand Plateaus: Capitalism and Schizophrenia*. London: Bloomsbury Publishing.

References

Deleuze, G. and Parnet, C. (2007). *Dialogues II*, rev. edn. New York: Columbia University Press.

Delgado, H. (1993). *New Immigrants, Old Unions: Organizing Undocumented Workers in Los Angeles*. Philadelphia: Temple University Press.

Diaz-Strong, D. X. and Ybarra, M. A. (2016). Disparities in High School Completion among Latinos: The Role of the Age-at-arrival and Immigration Status. *Children and Youth Services Review* 71: 282–9. https://doi.org/10.1016/j.childyouth.2016.11.021

Diken, B. (2004). From Refugee Camps to Gated Communities: Biopolitics and the End of the City. *Citizenship Studies* 8(1): 83–106. https://doi.org/10.1080/1362102042000178373

Dirty Pretty Things (2004). Larson, N., Frears, S., Tautou, A., Ejiofor, C., López, S., Menges, C., Knight, S., Miramax Home Entertainment (Firm). Miramax Films.

Donato, K. and Armenta, A. (2011). What We Know About Unauthorized Migration. *Annual Review of Sociology* 37: 529–43.

Donato, K., Tolbert II, C. M., Nucci, A., and Kawano, Y. (2007). Recent Immigration Settlement in the Nonmetropolitan United States: Evidence from Internal Census Data. *Rural Sociology* 72(4): 537–59.

Doty, R. and Wheatley, E. S. (2013). Private Detention and the Immigration Industrial Complex. *International Political Sociology* 7(4): 426–43. https://doi.org/10.1111/ips.12032

Douglas, K. M. and Sáenz, R. (2013). The Criminalization of Immigrants and the Immigration–Industrial Complex. *Daedalus* 142(3): 199–227. https://doi.org/10.1162/DAED_a_00228

Dreby, J. (2010). *Divided by Borders: Mexican Migrants and Their Children*. Berkeley, CA: University of California Press.

Dreby, J. (2015). *Everyday Illegal: When Policies Undermine Immigrant Families*. Berkeley, CA: University of California Press.

Dreby, J. and Schmalzbauer, L. (2013). The Relational Contexts of Migration: Mexican Women in New Destination Sites. *Sociological Forum* 28(1): 1–26. https://doi.org/10.1111/socf.12000

Dunn, T. (1996). *The Militarization of the US–Mexico Border, 1978–1992: Low-Intensity Conflict Doctrine Comes Home*. Austin, TX: University of Texas.

Dunn, T. J. (2001). Border Militarization via Drug and Immigration Enforcement: Human Rights Implications. *Social Justice* 28(2) (84): 7–30.

Düvell, F. (2008). Clandestine Migration in Europe. *Social Science Information* 47(4): 479–97. https://doi.org/10.1177/0539018408096442

Edkins, J. (2000). Sovereign Power, Zones of Indistinction, and the Camp. *Alternatives* 25(1): 3–25.

Ellis, B. D., Gonzales, R. G., and Rendón García, S. A. (2018). The Power of Inclusion: Theorizing "Abjectivity" and Agency under DACA. *Cultural Studies↔Critical Methodologies*, 1532708618817880.

References

Enriquez, L. E. and Saguy, A. C. (2016). Coming Out of the Shadows: Harnessing a Cultural Schema to Advance the Undocumented Immigrant Youth Movement. *American Journal of Cultural Sociology* 4(1): 107–30. https://doi.org/10.1057/ajcs.2015.6

Erikson, E. H. (1963). *Childhood and Society*. Harmondsworth: Penguin.

European Council (2016). EU–Turkey Statement. Press release, March 18. https://www.consilium.europa.eu/en/press/press-releases/2016/03/18/eu-turkey-statement/

Eurostat (2018). *The EU in the World*. Luxembourg: Publications Office of the European Union. https://ec.europa.eu/eurostat/documents/3217494/9066251/KS-EX-18-001-EN-N.pdf/64b85130-5de2-4c9b-aa5a-8881bf6ca59b.

Fekete, L. (2005). The Deportation Machine: Europe, Asylum and Human Rights. *Race & Class* 47(1): 64–78. https://doi.org/10.1177/0306396805055083

Fekete, L. (2007). Detained: Foreign Children in Europe. *Race & Class* 49(1): 93–104. https://doi.org/10.1177/0306396807080071

Feliciano, C. (2005). Does Selective Migration Matter? Explaining Ethnic Disparities in Educational Attainment among Immigrants' Children. *International Migration Review* 39(4): 841–71. https://doi.org/10.1111/j.1747-7379.2005.tb00291.x

Feliciano, C. (2006). Beyond the Family: The Influence of Premigration Group Status on the Educational Expectations of Immigrants' Children. *Sociology of Education* 79(4): 281–303. https://doi.org/10.1177/003804070607900401

Fenster, T. (2005). The Right to the Gendered City: Different Formations of Belonging in Everyday Life. *Journal of Gender Studies* 14(3): 217–31. doi: 10.1080/09589230500264109

Finotelli, C. and Arango, J. (2011). Regularisation of Unauthorised Immigrants in Italy and Spain: Determinants and Effects. *Documents d'anàlisi geogràfica* 57(3): 495–515.

Flores, W. V. and Benmayor, R. (1997). *Latino Cultural Citizenship: Claiming Identity, Space, and Rights*. Boston, MA: Beacon Press.

Flynn, D. (2005). New Borders, New Management: The Dilemmas of Modern Immigration Policies. *Ethnic and Racial Studies* 28(3): 463–90.

Fortuny, K., Capps, R., and Passel, J. (2007). *The Characteristics of Unauthorized Immigrants in California, Los Angeles County, and the United States*. Washington, DC: The Urban Institute. https://www.urban.org/sites/default/files/publication/46376/411425-The-Characteristics-of-Unauthorized-Immigrants-in-California-Los-Angeles-County-and-the-United-States.PDF

Foucault, M. (1979). *Discipline and Punish: The Birth of the Prison*. New York: Vintage Books. http://nrs.harvard.edu/urn-3:hul.ebookbatch.ASP_batch:ASPS100021788soth

Furuseth, O., Smith, H., and McDaniel, P. (2015). Belonging in Charlotte: Multiscalar Differences in Local Immigration Politics and Policies.

References

Geographical Review 105(1): 1–19. http://dx.doi.org.ezp-prod1.hul.harvard.edu/10.1111/j.1931-0846.2014.12048.x

Fussell, E. (2011). The Deportation Threat Dynamic and Victimization of Latino Migrants: Wage Theft and Robbery. *The Sociological Quarterly* 52(4): 593–615. https://doi.org/10.1111/j.1533-8525.2011.01221.x

Gálvez, A. (2013). Immigrant Citizenship: Neoliberalism, Immobility and the Vernacular Meanings of Citizenship. *Identities* 20(6): 1–18.

Gammeltoft-Hansen, T. and Sørensen, N. N. (2012). *The Migration Industry and the Commercialization of International Migration*. New York: Routledge.

Garcés-Mascareñas, B. (2018). "Ciutats refugi: una alternativa?" *Barcelona Metropolis* 109: 68–74.

García, M. C. (2006). *Seeking Refuge: Central American Immigration to Mexico, the United States, and Canada*. Berkeley, CA: University of California Press.

García Bedolla, L. (2005). *Fluid Borders: Latino Power, Identity, and Politics in Los Angeles*. Berkeley, CA: University of California Press.

Giroux, H. (1983). Theories of Reproduction and Resistance in the New Sociology of Education: A Critical Analysis. *Harvard Educational Review* 53(3): 257–93.

Gleeson, S. (2009). From Rights to Claims: The Role of Civil Society in Making Rights Real for Vulnerable Workers. *Law & Society Review* 43(3): 669–700.

Gleeson, S. (2010). Labor Rights for All? The Role of Undocumented Immigrant Status for Worker Claims Making. *Law and Social Inquiry Journal of the American Bar Foundation* 35(3): 561–602.

Gleeson, S. and Gonzales, R. G. (2012). When Do Papers Matter? An Institutional Analysis of Undocumented Life in the United States. *International Migration* 50(4): 1–19. https://doi.org/10.1111/j.1468-2435.2011.00726.x

Golash-Boza, T. (2009). The Immigration Industrial Complex: Why We Enforce Immigration Policies Destined to Fail. *Sociology Compass* 3(2): 295–309. https://doi.org/10.1111/j.1751-9020.2008.00193.x

Golash-Boza, T. (2012). *Due Process Denied: Detentions and Deportations in the United States*. London: Routledge.

Golash-Boza, T. M. (2015a). *Deported: Immigrant Policing, Disposable Labor, and Global Capitalism*. New York: New York University Press.

Golash-Boza, T. M. (2015b). *Immigration Nation: Raids, Detentions, and Deportations in Post-9/11 America*. Abingdon, UK: Routledge.

Golash-Boza, T. (2016). Feeling Like a Citizen, Living as a Denizen: Deportees' Sense of Belonging. *American Behavioral Scientist* 60(13): 1575–89. https://doi.org/10.1177/0002764216664943

Golash-Boza, T. M. (ed.). (2017). *Forced Out and Fenced In: Immigration Tales from the Field: A Collection of Essays*. New York: Oxford University Press.

Golash-Boza, T. and Hondagneu-Sotelo, P. (2013). Latino Immigrant Men and the Deportation Crisis: A Gendered Racial Removal Program. *Latino Studies* 11(3): 271–92. https://doi.org/10.1057/lst.2013.14

References

Goldring, L. and Landolt, P. (2012). The Impact of Precarious Legal Status on Immigrants' Economic Outcomes. *Institute for Research on Public Policy Study* 35, 1.

Goldring, L., Berinstein, C., and Bernhard, J. K. (2009). Institutionalizing Precarious Migratory Status in Canada. *Citizenship Studies* 13(3): 239–65. https://doi.org/10.1080/13621020902850643

Gomberg-Muñoz, R. (2011). *Labor and Legality: An Ethnography of a Mexican Immigrant Network*. New York: Oxford University Press.

Gomberg-Muñoz, R. (2017). *Becoming Legal: Immigration Law and Mixed-status Families*. New York: Oxford University Press.

Gonella, C. (2017). Visa Overstays Outnumber Illegal Border Crossings, Trends Expected to Continue. NBC News, March 7. https://www.nbcnews.com/news/latino/visa-overstays-outnumber-illegal-border-crossings-trend-expected-continue-n730216

Gonzales, R. G. (2008). Left Out But Not Shut Down: Political Activism and the Undocumented Student Movement. *Northwestern Journal of Law & Social Policy* 3: 219–63.

Gonzales, R. G. (2010a). More than Just Access: Undocumented Students Navigating the Postsecondary Terrain (report). *The Journal of College Admission*. https://vtechworks.lib.vt.edu/handle/10919/86901

Gonzales, R. G. (2010b). On the Wrong Side of the Tracks: Understanding the Effects of School Structure and Social Capital in the Educational Pursuits of Undocumented Immigrant Students. *Peabody Journal of Education* 85(4): 469–85. https://doi.org/10.1080/0161956X.2010.518039

Gonzales, R. G. (2011). Learning to Be Illegal: Undocumented Youth and Shifting Legal Contexts in the Transition to Adulthood. *American Sociological Review* 76(4): 602–19. https://doi.org/10.1177/0003122411411901

Gonzales, R. G. (2016). *Lives in Limbo: Undocumented and Coming of Age in America*. Berkeley, CA: University of California Press.

Gonzales, R. G. and Burciaga, E. M. (2018). Segmented Pathways of Illegality: Reconciling the Coexistence of Master and Auxiliary Statuses in the Experiences of 1.5-generation Undocumented Young Adults. *Ethnicities* 18(2): 178–91. https://doi.org/10.1177/1468796818767176

Gonzales, R. G. and Chavez, L. (2012). "Awakening to a Nightmare": Abjectivity and Illegality in the Lives of Undocumented 1.5-Generation Latino Immigrants in the United States. *Current Anthropology* 53(3): 255–81. https://doi.org/10.1086/665414

Gonzales, R. G. and Raphael, S. (2017). Illegality: A Contemporary Portrait of Immigration. *RSF: The Russell Sage Foundation Journal of the Social Sciences* 3(4): 1–17. https://doi.org/10.7758/rsf.2017.3.4.01

Gonzales, R. G. and Ruiz, A. (2014). Dreaming Beyond the Fields: Undocumented Youth, Rural Realities and a Constellation of Disadvantage. *Latino Studies* 12(2): 194–216.

References

Gonzales, R. G. and Sigona, N. (2006). *Within and Beyond Citizenship: Borders, Membership and Belonging*. Abingdon, UK: Taylor and Francis.

Gonzales, R. G. and Sigona, N. (2017). Mapping the Soft Borders of Citizenship: An Introduction, in R. G. Gonzales and N. Sigona (eds), *Within and Beyond Citizenship: Borders, Membership and Belonging*. Abingdon, UK: Routledge (BSA Sociological Futures Series), pp. 1–16.

Gonzales, R. G., Heredia, L. L., and Negrón-Gonzales, G. (2015). Untangling Plyler's Legacy: Undocumented Students, Schools, and Citizenship. *Harvard Educational Review* 85(3): 318–41. http://dx.doi.org.ezp-prod1.hul.harvard.edu/10.17763/0017-8055.85.3.318

Gonzales, R. G., Suárez-Orozco, C., and Dedios-Sanguineti, M. C. (2013). No Place to Belong: Contextualizing Concepts of Mental Health among Undocumented Immigrant Youth in the United States. *American Behavioral Scientist* 57(8): 1174–99. https://doi.org/10.1177/0002764213487349

Gonzales, R. G., Ellis, B., Rendón-García, S. A., and Brant, K. (2018). (Un)authorized Transitions: Illegality, DACA, and the Life Course. *Research in Human Development* 15(3–4): 345–59.

Gonzalez-Barrera, A. and Manuel-Krogstad, J. (2014). *US Deportations of Immigrants Reach Record High in 2013*. Pew Research Center. http://www.pewresearch.org/fact-tank/2014/10/02/u-s-deportations-of-immigrants-reach-record-high-in-2013/

Gordon, I., Scanlon, K., Travers, T. and Whitehead, C. (2009) Economic Impact on the London and UK Economy of an Earned Regularisation of Irregular Migrants to the UK. Greater London Authority.

Guiraudon, V. (2001). Weak Weapons of the Weak? Transnational Mobilization around Migration in the European Union, in D. Imig and S. Tarrow (eds), *Contentious Europeans: Protest and Politics in an Emerging Polity*. Lanham, MD: Rowman and Littlefield, pp. 163–83.

Gusfield, J. R. (1975). *Community: A Critical Response*. Oxford: Blackwell.

Hagan, J. M. (1994). *Deciding to be Legal: A Maya Community in Houston*. Philadelphia: Temple University Press.

Haggerty, K. D. and Ericson, R. V. (2000). The Surveillant Assemblage. *The British Journal of Sociology* 51: 605–22. doi: 10.1080/00071310020015280

Hall, M. and Lee, B. (2010). How Diverse are US Suburbs? *Urban Studies* 47(1): 3–28. https://doi.org/10.1177/0042098009346862

Hall, M. and Stringfield, J. (2014). Undocumented Migration and the Residential Segregation of Mexicans in New Destinations. *Social Science Research* 47: 61–78.

Hammar, T. (1990). *Democracy and the Nation State: Aliens, Denizens, and Citizens in a World of International Migration*. Aldershot, UK and Brookfield, VT: Avebury and Gower PubCo.

Hastings, A. (1998). Connecting Linguistic Structures and Social Practices: A Discursive Approach to Social Policy Analysis. *Journal of Social Policy* 27(2): 191–211.

References

Heidbrink, L. (2014). *Migrant Youth, Transnational Families, and the State: Care and Contested Interests*, 1st edn. Philadelphia: University of Pennsylvania Press.

Heller, C. and Jones, C. (2014). Eurosur: Saving Lives or Reinforcing Deadly Borders. *Statewatch Journal* 23(3/4): 9–11.

Heredia, J. I. (2016). Migrating to the City: Negotiating Gender and Race in Marie Arana's Lima Nights. *Hispania* 99(3): 459–70.

Heyman, J. (1995). Putting Power in the Anthropology of Bureaucracy: The Immigration and Naturalization Service at the Mexico–United States Border. *Current Anthropology* 36(2): 261–87.

Holliday, A. L. and Dwyer, R. E. (2009). Suburban Neighborhood Poverty in US Metropolitan Areas in 2000. *City & Community* 8(2): 155–76. https://doi.org/10.1111/j.1540-6040.2009.01278.x

Holloway, J. (1994). Global Capital and the National State. *Capital & Class* 18(1): 23–49.

Holloway, J. (1995). From Scream of Refusal to Scream of Power: The Centrality of Work, in Werner Bonefeld, Richard Gunn, and Kosmas Psychopedis (eds), *Open Marxism*. London: Pluto Press, pp. 3–155.

Holmes, S. M. (2013). *Fresh Fruit, Broken Bodies: Migrant Farmworkers in the United States*. Berkeley, CA: University of California Press.

Home Affairs Select Committee (2013). The Work of the UK Border Agency (July–September 2013). London: Home Office.

Hondagneu-Sotelo, P. (1994). *Gendered Transitions: Mexican Experiences of Immigration*. Berkeley, CA: University of California Press.

Horton, S. (2009). A Mother's Heart is Weighed Down with Stones: A Phenomenological Approach to the Experience of Transnational Motherhood. *Culture, Medicine and Psychiatry* 33(1): 21–40. http://dx.doi.org.ezp-prod1.hul.harvard.edu/10.1007/s11013-008-9117-z

Horton, S. B. (2016). *They Leave Their Kidneys in the Fields: Illness, Injury, and Illegality among US Farmworkers*. Berkeley, CA: University of California Press. http://nrs.harvard.edu/urn-3:hul.ebookbatch.PMUSE_batch:20170202muse53083

Hughes, E. C. (1945). Dilemmas and Contradictions of Status. *American Journal of Sociology* 50(5): 353–9.

Humphris, R, and Sigona, N. (2016) Children and Unsafe Migration in Europe: Data and Policy – Understanding the Evidence Base. IOM GMDAC Data Briefing, 5.

ICE (Immigration and Customs Enforcement) (2018a). Delegation of Immigration Authority Section 287(g) Immigration and Nationality Act. https://www.ice.gov/287g

ICE (Immigration and Customs Enforcement) (2018b). Secure Communities. https://www.ice.gov/secure-communities

ICE (n.d.). Directive Identification and Monitoring of Pregnant Detainees.

References

https://www.ice.gov/directive-identification-and-monitoring-pregnant-detainees

ICIBI (Independent Chief Inspector of Borders and Immigration) (2015). ICIBI Report on Illegal Working, December 2015. https://www.gov.uk/government/publications/inspection-report-on-illegal-working-december-2015

IDOS Study and Research Centre/Statistical Dossier on Immigration (2018). Compiled Data for 2018. https://www.dossierimmigrazione.it/appuntamenti.php?m=&y=2018

Inda, J. X. (2006). *Targeting Immigrants: Government, Technology, and Ethics.* Malden, MA: Blackwell Publishing.

IOM (International Organization for Migration) (2019). World Migration Report. Geneva: UN https://publications.iom.int/system/files/pdf/wmr_2018_en.pdf

Isin, E. F. (2002). *Being Political: Genealogies of Citizenship.* Minneapolis, MN: University of Minnesota Press.

Isin, E. F. (2008). Theorizing Acts of Citizenship, in E. F. Isin and G. M. Nielsen (eds), *Acts of Citizenship* (pp. 15–43). London, UK: Palgrave Macmillan. http://zedbooks.co.uk/hardback/acts-of-citizenship

Isin, E. F. and Rygiel, K. (2007). Abject Spaces: Frontiers, Zones, Camps, in E. Dauphinee and C. Masters (eds), *The Logics of Biopower and the War on Terror: Living, Dying, Surviving* (pp. 181–203). New York: Palgrave Macmillan. https://doi.org/10.1007/978-1-137-04379-5_9

Jablon, R. and Thanawala, S. (2018). Judge Blocks US from Ending Protections for Some Immigrants. Associated Press, October 3, 2018. https://apnews.com/d4da7edc4258437eab021f6b8cdf2f8d

Jasso, G., Massey, D., Rosenzweig, M., and Smith, J. (2008). From Illegal to Legal: Estimating Previous Illegal Experience among New Legal Immigrants to the United States. *International Migration Review* 42(4): 803–43.

Johnson, J. C. (2014). Secure Communities Memo from Department of Homeland Security https://www.dhs.gov/sites/default/files/publications/14_1120_memo_secure_communities.pdf

Johnson, K. R. (2011). Sweet Home Alabama: Immigration and Civil Rights in the New South. *Stanford Law Review Online* 64: 22–8.

Jones, H., Gunaratnam, Y., Bhattacharyya, G., Davies, W., et al. (2017). *Go Home? The Politics of Immigration Controversies.* Manchester, UK: Manchester University Press.

Jones, R. (2016). *Violent Borders: Refugees and the Right to Move.* London and New York: Verso Books.

Jones-Correa, M. (2005, February). The bureaucratic incorporation of immigrants in suburbia, in *Immigration to the United States: New Sources and Destinations Conference.* New York: Russell Sage Foundation, pp. 3–4.

Jones-Correa, M. and de Graauw, E. (2013). The Illegality Trap: The Politics of

References

Immigration and the Lens of Illegality. *Daedalus* 142(3): 185–98. https://doi.org/10.1162/DAED_a_00227

Joppke, C. (1998). Why Liberal States Accept Unwanted Immigration. *World Politics* 50(2): 266.

Jordan, M. (2018, December 25). 8-Year-Old Migrant Child from Guatemala Dies in US Custody. *New York Times*. https://www.nytimes.com/2018/12/25/us/guatemalan-boy-dies-border-patrol.html

Jusionyte, I. (2018). *Threshold: Emergency Responders on the US–Mexico Border*. Berkeley, CA: University of California Press.

Kandel, W., Henderson, J., Koball, H., and Capps, R. (2011). Moving Up in Rural America: Economic Attainment of Nonmetro Latino Immigrants. *Rural Sociology* 76(1): 101–28. https://doi.org/10.1111/j.1549-0831.2011.00047.x

Katz, V. (2014). Children as Brokers of Their Immigrant Families' Health-Care Connections. *Social Problems* 61(2): 194–215. https://doi.org/10.1525/sp.2014.12026

Keshavarz, M. and Zetterlund, C. (2013). A Method for Materialising Borders, in *Silent University Reader*. Tensta Konsthall/Silent University. http://thesilentuniversity.org/

Krogstad, J. M., Passel, J. S., and Cohn, D. (2018). Five Facts about Illegal Immigration in the US. Washington, DC: Pew Research Center. https://www.pewhispanic.org/2014/09/03/as-growth-stalls-unauthorized-immigrant-population-becomes-more-settled/

Kubal, A. (2013). Conceptualizing Semi-Legality in Migration Research. *Law & Society Review* 47(3): 555–87. https://doi.org/10.1111/lasr.12031

Lahav, G. and Guiraudon, V. (2006). Actors and Venues in Immigration Control: Closing the Gap between Political Demands and Policy Outcomes. *West European Politics* 29(2): 201–23. https://doi.org/10.1080/01402380500512551

Landale, N. S., Oropesa, R. S., and Gorman, B. K. (2000). Migration and Infant Death: Assimilation or Selective Migration among Puerto Ricans? *American Sociological Review* 65(6): 888–909.

Lee, J. and Zhou, M. (2015). *The Asian American Achievement Paradox*. New York: Russell Sage Foundation. http://nrs.harvard.edu/urn-3:hul.ebookbatch.PMUSE_batch:muse4743920151126

Lee, S. (2015). *Growing Up Outside the Law*. Harvard Law Review. https://harvardlawreview.org/2015/03/growing-up-outside-the-law/

Leerkes, A., Bachmeier, J. D., and Leach, M. A. (2013). When the Border Is "Everywhere": State-level Variation in Migration Control and Changing Settlement Patterns of the Unauthorized Immigrant Population in the United States. *International Migration Review* 47(4): 910–43. https://doi.org/10.1111/imre.12047

Lehman, B. (2016). Latino Students in New Destinations: Immigration, Extracurricular Activities, and Bullying Victimization. *Education and Youth Today* 20(1): 123–44.

References

Lewis, P. and Ramakrishnan, S. K. (2007). Police Practices in Immigrant-Destination Cities: Political Control or Bureaucratic Professionalism? *Urban Affairs Review* 42(6): 874–900. https://doi.org/10.1177/1078087407300752

Lichter, D. T. and Johnson, K. M. (2007). The Changing Spatial Concentration of America's Rural Poor Population. *Rural Sociology* 72(3): 331–58. http://dx.doi.org.ezp-prod1.hul.harvard.edu/10.1526/003601107781799290

Lichter, D. T. and Johnson, K. M. (2009). Immigrant Gateways and Hispanic Migration to New Destinations. *International Migration Review* 43(3): 496–518.

Lichter, D. T., Parisi, D., Taquino, M. C., and Grice, S. M. (2010). Residential Segregation in New Hispanic Destinations: Cities, Suburbs, and Rural Communities Compared. *Social Science Research* 39(2): 215–230. https://doi.org/10.1016/j.ssresearch.2009.08.006

Light, I. (2006). International Migration: Prospects and Policies in a Global Market. *Contemporary Sociology* 35(1): 49–51. https://doi.org/10.1177/009430610603500135

Lipsky, M. (2010). *Street-Level Bureaucracy, 30th Ann. Ed.: Dilemmas of the Individual in Public Service*. New York: Russell Sage Foundation.

López, J. L. (2015). "Impossible Families": Mixed-Citizenship Status Couples and the Law. *Law & Policy* 37(1/2): 93–118. https://doi.org/10.1111/lapo.12032

Macías-Rojas, P. (2016). *From Deportation to Prison: The Politics of Immigration Enforcement in Post-Civil Rights America*. New York: New York University Press.

Maestri, G., and Hughes, S. M. (2017). Contested Spaces of Citizenship: Camps, Borders and Urban Encounters. *Citizenship Studies* 21(6): 625–39. https://doi.org/10.1080/13621025.2017.1341657

Marrow, H. B. (2009). Immigrant Bureaucratic Incorporation: The Dual Roles of Professional Missions and Government Policies. *American Sociological Review* 74(5): 756–76. https://doi.org/10.1177/000312240907400504

Marrow, H. B. (2011). *New Destination Dreaming: Immigration, Race, and Legal Status in the Rural American South*. Redwood City, CA: Stanford University Press.

Marshall, T. H. (1950). *Citizenship and Social Class and Other Essays*. Cambridge: Cambridge University Press.

Martin, P. L., Cornelius, W. A., and Hollifield, J. F. (eds). (1994). *Controlling Immigration: A Global Perspective*. Redwood City, CA: Stanford University Press.

Massey, D. S. (1999). International Migration at the Dawn of the Twenty-First Century: The Role of the State. *Population and Development Review* 25(2): 303–22. https://doi.org/10.1111/j.1728-4457.1999.00303.x

Massey, D. S. (2008). *New Faces in New Places: The Changing Geography of American Immigration*. New York: Russell Sage Foundation. http://nrs.harvard.edu/urn-3:hul.ebookbatch.PMUSE_batch:muse9781610443814

References

Massey, D. and Espinosa, K. (1997). What's Driving Mexico–US Migration? A Theoretical, Empirical, and Policy Analysis. *American Journal of Sociology* 102(4): 939–99. http://www.jstor.org/stable/2782024

Massey, D. S., Durand, J., and Malone, N. J. (2002). *Beyond Smoke and Mirrors: Mexican Immigration in an Era of Economic Integration*. New York: Russell Sage Foundation.

Massey, D. S., Alarcón, R., Durand, J., and González, H. (1990). *Return to Aztlan: The Social Process of International Migration from Western Mexico*. Berkeley, CA: University of California Press.

Massey, D. S., Arango, J., Hugo, G., Kouaouci, A., Pellegrino, A., and Taylor, J. E. (1993). Theories of International Migration: A Review and Appraisal. *Population and Development Review* 19(3): 431–66.

Massey, D. S., Arango, J., Hugo, G., Kouaouci, A., Pellegrino, A., and Taylor, J. (1994). An Evaluation of International Migration Theory: The North American Case. *Population and Development Review* 20(4): 699–751.

Mateo, L. (2013, July 22). The Fight to Keep Families Together Does Not End at Deportation. Retrieved February 2, 2019, from https://www.huffingtonpost.com/lizbeth-mateo/the-fight-to-keep-familie_b_3634915.html

McMahon, S. and Sigona, N. (2018). Navigating the Central Mediterranean in a Time of "Crisis": Disentangling Migration Governance and Migrant Journeys. *Sociology* 52(3): 497–514. https://doi.org/10.1177/0038038518762082

McMillan, D. W. and Chavis, D. M. (1986). Sense of Community: A Definition and Theory. *Journal of Community Psychology* 14(1): 6–23. doi: 10.1002/1520-6629(198601)14:1<6::AID-JCOP2290140103>3.0.CO;2-I

McNevin, A. (2013). Ambivalence and Citizenship: Theorising the Political Claims of Irregular Migrants. *Millennium* 41(2): 182–200. https://doi.org/10.1177/0305829812463473

Meissner, D., Kerwin, D., Chishti, M., and Bergeron, C. (2013). Immigration Enforcement in the United States: The Rise of a Formidable Machinery. Migration Policy Institute. http://search.proquest.com/docview/1820726103/?pq-origsite=primo

Menjívar, C. (2000). *Fragmented Ties: Salvadoran Immigrant Networks in America*. Berkeley, CA: University of California Press.

Menjívar, C. (2006). Liminal Legality: Salvadoran and Guatemalan Immigrants' Lives in the United States. *American Journal of Sociology* 111(4): 999–1037. https://doi.org/10.1086/499509

Menjívar, C. (2016). Immigrant Criminalization in Law and the Media: Effects on Latino Immigrant Workers' Identities in Arizona. *American Behavioral Scientist* 60(5–6), 597–616. https://doi.org/10.1177/0002764216632836

Menjívar, C. and Abrego, L. (2009). Parents and Children across Borders: Legal Instability and Intergenerational Relations in Guatemalan and Salvadoran Families, in N. Foner (ed.), *Across Generations: Immigrant Families in America*. New York: New York University Press, pp. 160–89.

References

Menjívar, C. and Abrego, L. (2012). Legal Violence: Immigration Law and the Lives of Central American Immigrants. *American Journal of Sociology* 117(5): 1380–1421. https://doi.org/10.1086/663575

Menjívar, C. and Kanstroom, D. (2013). *Constructing Immigrant "Illegality": Critiques, Experiences, and Responses*. New York: Cambridge University Press.

Menjívar, C., Abrego, L. J., and Schmalzbauer, L. C. (2016). *Immigrant Families*. Cambridge: Polity Press.

Mezzadra, S. (2004). The Right to Escape. *Ephemera* 4(3): 267–75.

Mezzadra, S. and Neilson, B. (2013). *Border as Method, or, the Multiplication of Labor*. Durham, NC: Duke University Press.

Milkman, R. (2011). Immigrant Workers, Precarious Work, and the US Labor Movement. *Globalizations* 8(3): 361–72. https://doi.org/10.1080/14747731.2011.576857

Mitropoulos, A. and Neilson, B. (2006). Exceptional Times, Non-governmental Spacings, and Impolitical Movements – Vacarme. https://vacarme.org/article484.html

Moffette, D. (2018). The Jurisdictional Games of Immigration Policing: Barcelona's Fight against Unauthorized Street Vending. *Theoretical Criminology*. https://doi.org/10.1177/1362480618811693

Molina, N. (2014). *How Race is Made in America: Immigration, Citizenship, and the Historical Power of Racial Scripts*. Berkeley, CA: University of California Press.

Mountz, A. (2012). Mapping Remote Detention: Dis/location through Isolation, in J. Loyd, M. Mitchelson, and A. Burridge (eds), *Beyond Walls and Cages: Prisons, Borders, and Global Crisis*. Athens, GA: University of Georgia Press, pp. 91–104.

Mountz, A., Coddington, K., Loyd, J., and Catania, R. T. (2012) Conceptualizing Detention: Mobility, Containment, Bordering, and Exclusion. *Progress in Human Geography* 37(4): 522–41.

Nail, T. (2015). *The Figure of the Migrant*. Redwood City, CA: Stanford University Press.

Negrón-Gonzales, G. (2013). Navigating "Illegality": Undocumented Youth and Oppositional Consciousness. *Children and Youth Services Review* 35(8): 1284–90. http://dx.doi.org/10.1016/j.childyouth.2013.04.016

Negrón-Gonzales, G. (2014). Undocumented, Unafraid and Unapologetic: Re-articulatory Practices and Migrant Youth "Illegality." *Latino Studies* 12(2): 259–78. http://dx.doi.org.ezp-prod1.hul.harvard.edu/10.1057/lst.2014.20

Nevins, J. (2002). *Operation Gatekeeper: The Rise of the "Illegal Alien" and the Making of the US–Mexico Boundary*. New York: Routledge.

Ngai, M. M. (2004). *Impossible Subjects: Illegal Aliens and the Making of Modern America*. Princeton: Princeton University Press.

Nicholls, W. (2013). *The DREAMers: How the Undocumented Youth Movement*

References

Transformed the Immigrant Rights Debate. Redwood City, CA: Stanford University Press.

Nicholls, W. (2016). Politicizing Undocumented Immigrants One Corner at a Time: How Day Laborers Became a Politically Contentious Group. *International Journal of Urban and Regional Research* 40(2): 299–320. https://doi.org/10.1111/1468-2427.12334

Nicholls, W. J. and Fiorito, T. (2015, January). Dreamers Unbound: Immigrant Youth Mobilizing. *New Labor Forum* 24(1): 86–92. https://doi.org/10.1177/1095796014562234

Nicholls, W. J., Maussen, M., and de Mesquita, L. C. (2016). The Politics of Deservingness: Comparing Youth-Centered Immigrant Mobilizations in the Netherlands and the United States. *American Behavioral Scientist* 60(13): 1590–1612.

Nyers, P. and Rygiel, K. (2012). *Citizenship, Migrant Activism and the Politics of Movement*. Abingdon, UK: Routledge.

Okamoto, D. and Ebert, K. (2010). Beyond the Ballot: Immigrant Collective Action in Gateways and New Destinations in the United States. *Social Problems* 57(4): 529–58. https://doi.org/10.1525/sp.2010.57.4.529

Olivas, M. (2012). Dreams Deferred: Deferred Action, Prosecutorial Discretion, and the Vexing Case(s) of Dream Act Students. *The William and Mary Bill of Rights Journal* 21(2): 463–547.

Ortega, A. N., Horwitz, S. M., Fang, H., Kuo, A. A. et al. (2009). Documentation Status and Parental Concerns about Development in Young US Children of Mexican Origin. *Academic Pediatrics* 9(4): 278–82. https://doi.org/10.1016/j.acap.2009.02.007

Osgood, D. W., Foster, E. M., Flanagan, C., and Ruth, G. R. (2005). *On Your Own without a Net: The Transition to Adulthood for Vulnerable Populations*. Chicago, IL: University of Chicago Press.

Papadopoulos, R. (2008). Systematic Challenges in a Refugee Camp. *Context* 99: 16–19.

Papastergiadis, N. (2006). The Invasion Complex: The Subject Other and Spaces of Violence. *Geografiska Annaler: Series B, Human Geography* 88(4): 429–42.

Papoutsi, A., Painter, J., Papada, E., and Vradis, A. (2018). The EC Hotspot Approach in Greece: Creating Liminal EU Territory. *Journal of Ethnic and Migration Studies*: 1–13.

Parisi, R. (2017). Squatting as a Practice of Citizenship: The Experiences of Moroccan Immigrant Women in Rome, in R. G. Gonzales and N. Sigona (eds), *Within and Beyond Citizenship: Borders, Membership and Belonging*. Abingdon, UK and New York: Routledge, ch. 7.

Parrado, E. A. and Flippen, C. A. (2016). The Departed: Deportations and Out-Migration among Latino Immigrants in North Carolina after the Great Recession. *The ANNALS of the American Academy of Political and Social Science* 666(1): 131–47. https://doi.org/10.1177/0002716216646563

References

Passel, J. and Cohn, D. (2018). US Unauthorized Immigrant Total Dips to Lowest Level in a Decade. Pew Research Center, November 27, 2018. https://www.pewhispanic.org/2018/11/27/u-s-unauthorized-immigrant-total-dips-to-lowest-level-in-a-decade/

Passel, J. and Lopez, M. (2012). *Up to 1.7 Million Unauthorized Immigrant Youth May Benefit from New Deportation Rules*. Washington, DC: Pew Research Center.

Pastor, M. and Mollenkopf, J. (2012). Struggling over Strangers or Receiving with Resilience? The Metropolitics of Immigrant Integration. Brookings Institution Press. https://www.jstor.org/stable/10.7864/j.ctt12632t

Portes, A. (1978). Migration and Underdevelopment. *Politics & Society* 8(1): 1–48. https://doi.org/10.1177/003232927800800101

Portes, A. and Rumbaut, R. G. (2001). *Legacies: The Story of the Immigrant Second Generation*. Berkeley, CA: University of California Press.

Portes, A. and Rumbaut, R. (2006). *Immigrant America: A Portrait* (3rd edn, rev., expanded, and updated). Berkeley, CA: University of California Press.

Portes, A. and Stepick, A. (1993). *City on the Edge: The Transformation of Miami*. Berkeley, CA: University of California Press.

Portes, A. and Zhou, M. (1993). The New Second Generation: Segmented Assimilation and its Variants. *The ANNALS of the American Academy of Political and Social Science* 530(1): 74–96. https://doi.org/10.1177/0002716293530001006

Preston, J. (2010, May 17). Illegal Immigrant Students Protest at McCain Office – New York Times. https://www.nytimes.com/2010/05/18/us/18dream.html

Probyn, E. (1996). *Outside Belongings*. New York: Routledge.

Provine, D. M., Varsanyi, M. W., Lewis, P. G., and Decker, S. H. (2016). *Policing Immigrants: Law Enforcement on the Front Line*. Chicago, IL: The University of Chicago Press.

Pruitt, L. R. (2007). Toward a Feminist Theory of the Rural. *Utah Law Review*, 421.

Purser, G. (2009). The Dignity of Job-Seeking Men: Boundary Work among Immigrant Day Laborers. *Journal of Contemporary Ethnography* 38(1): 117–39. https://doi.org/10.1177/0891241607311867

Ramadan, A. (2008). The Guests' Guests: Palestinian Refugees, Lebanese Civilians, and the War of 2006. *Antipode* 40(4): 658–77.

Ramadan, A. (2013). Spatialising the Refugee Camp. *Transactions of the Institute of British Geographers* 38(1): 65–77. https://doi.org/10.1111/j.1475-5661.2012.00509.x

Redclift, V. (2013). Abjects or Agents? Camps, Contests and the Creation of "Political Space." *Citizenship Studies* 17(3–4): 308–21. https://doi.org/10.1080/13621025.2013.791534

Redfield, P. (2005). Doctors, Borders, and Life in Crisis. *Cultural Anthropology* 20(3): 328–61. https://doi.org/10.1525/can.2005.20.3.328

References

Reed-Danahay, D. and Brettell, C. (2008). *Citizenship, Political Engagement, and Belonging: Immigrants in Europe and the United States*. New Brunswick, NJ: Rutgers University Press.

Reeves, M. (2015). Living from the Nerves: Deportability, Indeterminacy and the Feel of Law in Migrant Moscow. *Social Analysis* 59(4): 119–36.

Ridgley, J. (2008). Cities of Refuge: Immigration Enforcement, Police, and the Insurgent Genealogies of Citizenship in US Sanctuary Cities. *Urban Geography* 29(1): 53–77. https://doi.org/10.2747/0272-3638.29.1.53

Rincón, A. (2008). *Undocumented Immigrants and Higher Education: Sí se Puede!* New York: LFB Scholarly Pub. http://nrs.harvard.edu/urn-3:hul.ebook batch.EBRARY_batch:ebr10256710

Robinson, C. and Preston, J. (2012). Appeals Court Draws Boundaries on Alabama's Immigration Law. *New York Times*, August 21.

Rodríguez, E. G. (2004). "We Need your Support, but the Struggle is Primarily Ours": On Representation, Migration and the *Sans Papiers* Movement, ESF Paris, 12–15 November 2003. *Feminist Review* 77(1): 152–6.

Rodriguez, N. P. (1987). Undocumented Central Americans in Houston: Diverse Populations. *International Migration Review* 21(1): 4–26. https://doi.org/10.1177/019791838702100101

Rodriguez, N. (1996). The Battle for the Border: Notes on Autonomous Migration, Transnational Communities, and the State. *Social Justice* 23: 21–37.

Rosaldo, R. (1994). Cultural Citizenship and Educational Democracy. *Cultural Anthropology* 9(3): 402–11. https://doi.org/10.1525/can.1994.9.3.02a00110

Rosaldo, R. (1997). Cultural Citizenship, Inequality, and Multiculturalism, in William V. Flores and Rina Benmayor (eds), *Latino Cultural Citizenship: Claiming Identity, Space, and Rights*. Boston: Beacon Press.

Rosaldo, R., and Flores, W. V. (1997). Identity, Conflict, and Evolving Latino Communities: Cultural Citizenship in San Jose, California, in William V. Flores and Rina Benmayor (eds), *Latino Cultural Citizenship: Claiming Identity, Space, and Rights*. Boston: Beacon Press, pp. 57–96.

Rosas, G. (2012). *Barrio Libre: Criminalizing States and Delinquent Refusals of the New Frontier*. Durham and London: Duke University Press.

Rosenblum, M. R. and Ruiz Soto, A. G. (2015). "An Analysis of Unauthorized Immigrants in the United States by Country and Region of Birth," Migration Policy Institute, August 2015. https://www.migrationpolicy.org/research/analysis-unauthorized-immigrants-united-states-country-and-region-birth

Rosenblum, M. R., Meissner, D., Bergeron, C., and Hipsman, F. (2014). *The Deportation Dilemma: Reconciling Tough and Humane Enforcement*. Washington, DC: Migration Policy Institute.

Roth, B. J. (2015). Making Connections in Middle-Class Suburbs? Low-Income Mexican Immigrant Young Men, Social Capital, and Pathways to High School Graduation. *Journal of Immigrant & Refugee Studies* 13(4): 321–38.

Roth, B., Gonzales, R. G., and Lesniewski, J. (2015). Building a Stronger

References

Safety Net: Local Organizations and the Challenges of Serving Immigrants in the Suburbs. *Human Service Organizations Management Leadership & Governance* 39(4): 348–61. https://doi.org/10.1080/23303131.2015.1050143

Ruhs, M. and Anderson, B. (eds) (2010). *Who Needs Migrant Workers? Labour Shortages, Immigration, and Public Policy*. Oxford: Oxford University Press.

Rumbaut, R. G. (1997). Assimilation and its Discontents: Between Rhetoric and Reality. *International Migration Review* 31(4): 923–60.

Rygiel, K. (2011). Bordering Solidarities: Migrant Activism and the Politics of Movement and Camps at Calais. *Citizenship Studies* 15(1): 1–19. https://doi.org/10.1080/13621025.2011.534911

Rytter, M. (2012). Semilegal Family Life: Pakistani Couples in the Borderlands of Denmark and Sweden. *Global Networks* 12(1): 91–108.

Salter, M. B. (2013). To Make Move and Let Stop: Mobility and the Assemblage of Circulation. *Mobilities* 8(1): 7–19. doi: 10.1080/17450101.2012.747779

Santos, F. (2012). Arizona Immigration Law Survives Ruling. *The New York Times*, September 6. https://www.nytimes.com/2012/09/07/us/key-element-of-arizona-immigration-law-survives-ruling.html

Sarason, S. B. (1974). *The Psychological Sense of Community: Prospects for a Community Psychology*, 1st edn. San Francisco: Jossey-Bass.

Sassen, S. (1991). *The Global City: New York, London, Tokyo*, 2nd edn. Princeton, NJ: Princeton University Press.

Sassen, S. (1996). *Losing Control? Sovereignty in an Age of Globalization* (University seminars/Leonard Hastings Schoff memorial lectures). New York: Columbia University Press.

Sassen, S. (2000). Territory and Territoriality in the Global Economy. *International Sociology* 15(2): 372–93. https://doi.org/10.1177/0268580900015002014

Sassen, S. (2006). *Territory, Authority, Rights: From Medieval to Global Assemblages*. Princeton, NJ: Princeton University Press.

Schiller, N. G. and Çağlar, A. (2009). Towards a Comparative Theory of Locality in Migration Studies: Migrant Incorporation and City Scale. *Journal of Ethnic and Migration Studies* 35(2): 177–202.

Schmalzbauer, L. (2014). *The Last Best Place? Gender, Family, and Migration in the New West*. Redwood City, CA: Stanford University Press.

Schmidt, L. A. and Buechler, S. (2017). "I Risk Everything because I have Already Lost Everything": Central American Female Migrants Speak Out on the Migrant Trail in Oaxaca, Mexico. *Journal of Latin American Geography* 16(1): 139–64.

Schuster, L. (2005). A Sledgehammer to Crack a Nut: Deportation, Detention and Dispersal in Europe. *Social Policy & Administration* 39(6): 606–21.

Schuster, L. (2011). Turning Refugees into "Illegal Migrants": Afghan Asylum Seekers in Europe. *Ethnic and Racial Studies* 34(8): 1392–1407.

Seif, H. (2004). "Wise Up!" Undocumented Latino Youth, Mexican-American

References

Legislators, and the Struggle for Higher Education Access. *Latino Studies* 2(2): 210–30. https://doi.org/10.1057/palgrave.lst.8600080

Semple, K. (2018, April 24). Inside an Immigrant Caravan: Women and Children, Fleeing Violence. *The New York Times*. https://www.nytimes.com/2018/04/04/world/americas/mexico-trump-caravan.html

Semple, K. and Jordan, M. (2018, May 1). Migrant Caravan of Asylum Seekers Reaches US Border. *New York Times*. https://www.nytimes.com/2018/04/29/world/americas/mexico-caravan-trump.html

Sigona, N. (2012). "I Have Too Much Baggage": The Impacts of Legal Status on the Social Worlds of Irregular Migrants. *Social Anthropology* 20(1): 50–65. https://doi.org/10.1111/j.1469-8676.2011.00191.x

Sigona, N. (2015). Campzenship: Reimagining the Camp as a Social and Political Space. *Citizenship Studies* 19(1): 1–15. https://doi.org/10.1080/13621025.2014.937643

Sigona, N. (2018). The Contested Politics of Naming in Europe's "Refugee Crisis." *Ethnic and Racial Studies* 41(3): 456–60. https://doi.org/10.1080/01419870.2018.1388423

Sigona, N. and Hughes, V. (2012). No Way Out, No Way In: Irregular Migrant Children and Families in the UK – COMPAS. https://www.compas.ox.ac.uk/2012/pr-2012-undocumented_migrant_children/

Sigona, N., Chase, E., and Humphris, R. (2017a) Understanding Causes and Consequences of Going "Missing," *Becoming Adult Research Brief No. 6*. London: UCL.

Sigona, N., Chase, E., and Humphris, R. (2017b) Protecting the 'Best Interests' of the Child in Transition to Adulthood. *Becoming Adult Research Brief No. 3*. London: UCL.

Silbey, S. S. (2005). After Legal Consciousness. *Annual Review of Law and Social Science* 1(1): 323–68. https://doi.org/10.1146/annurev.lawsocsci.1.041604.115938

Silverman, S. J. (2012). "Regrettable but Necessary?" A Historical and Theoretical Study of the Rise of the UK Immigration Detention Estate and its Opposition. *Politics & Policy* 40(6): 1131–57.

Simmons, W. P., Menjívar, C., and Téllez, M. (2015). Violence and Vulnerability of Female Migrants in Drop Houses in Arizona: The Predictable Outcome of a Chain Reaction of Violence. *Violence Against Women* 21(5), 551–70. https://doi.org/10.1177/1077801215573331

Simpson, J. and Weiner, E. S. (1989). *Oxford English Dictionary* online. Oxford: Clarendon Press.

Singer, A. (2004). The Rise of New Immigrant Gateways. *Brookings Institution Reports*. Brookings Institution Reports, February.

Slack, J., Martínez, D. E., Lee, A. E., and Whiteford, S. (2016). The Geography of Border Militarization: Violence, Death and Health in Mexico and the United States. *Journal of Latin American Geography* 15(1): 7–32.

References

South Carolina Community Loan Fund (2013). Rural South Carolina Faces Unprecedented Challenges. December 4.

Soysal, Y. N. (1994). *Limits of Citizenship: Migrants and Postnational Membership in Europe*. Chicago: University of Chicago Press.

Spencer, S. (2016). Postcode Lottery for Europe's Undocumented Children: Unravelling an Uneven Geography of Entitlements in the European Union. *American Behavioral Scientist* 60(13): 1613–28. https://doi.org/10.1177/0002764216664945

Speri, A. (2018, April 11). Detained, Then Violated: 1,224 Complaints Reveal a Staggering Pattern of Sexual Abuse in Immigration Detention. Half of Those Accused Worked for ICE. https://theintercept.com/2018/04/11/immigration-detention-sexual-abuse-ice-dhs/

Stein, G. L., Gonzales, R. G., Coll, C. G., and Prandoni, J. I. (2016). Latinos in Rural, New Immigrant Destinations: A Modification of the Integrative Model of Child Development, in L. J. Crockett and G. Carlo (eds), *Rural Ethnic Minority Youth and Families in the United States*. Cham: Springer, pp. 37–56.

Stephen, L. (2003). Cultural Citizenship and Labor Rights for Oregon Farmworkers: The Case of Pineros y Campesinos Unidos del Nordoeste (PCUN). *Human Organization* 62(1): 27–38.

Street, A., Jones-Correa, M., and Zepeda-Millán, C. (2017). Political Effects of Having Undocumented Parents. *Political Research Quarterly* 70(4): 818–32. https://doi.org/10.1177/1065912917717351

Stuesse, A. (2016). *Scratching Out a Living: Latinos, Race, and Work in the Deep South*. Berkeley, CA: University of California Press.

Suárez-Orozco, C., Bang, H. J., and Kim, H. Y. (2010). I Felt Like My Heart Was Staying Behind: Psychological Implications of Family Separations and Reunifications for Immigrant Youth. *Journal of Adolescent Research* 26(2): 222–57. https://doi.org/10.1177/0743558410376830

Suárez-Orozco, C., Pimentel, A., and Martin, M. (2009). The Significance of Relationships: Academic Engagement and Achievement among Newcomer Immigrant Youth. *Teachers College Record* 111(3): 712–49.

Suárez-Orozco, C., Suárez-Orozco, M. M., and Todorova, I. (2008). *Learning a New Land: Immigrant Students in American Society*. Cambridge, MA: Harvard University Press.

Suárez-Orozco, C., Yoshikawa, H., Teranishi, R. T., and Suárez-Orozco, M. M. (2011). Growing Up in the Shadows: The Developmental Implications of Unauthorized Status. *Harvard Educational Review* 81(3): 438–73. https://doi.org/10.17763/haer.81.3.g23x203763783m75

Swerts, T. (2017) Marching Beyond Borders: Non-Citizen Citizenship and Transnational Undocumented Activism in Europe, in R. G. Gonzales and N. Sigona (eds), *Within and Beyond Citizenship: Borders, Membership and Belonging*. Abingdon, UK: Taylor and Francis, ch. 9.

Terrio, S. J. (2015). *Whose Child am I? Unaccompanied, Undocumented*

References

Children in US Immigration Custody. Berkeley, CA: University of California Press.

Terriquez, V. and Kwon, H. (2015). Intergenerational Family Relations, Civic Organisations, and the Political Socialisation of Second-Generation Immigrant Youth. *Journal of Ethnic and Migration Studies* 41(3): 425–47. https://doi.org/10.1080/1369183X.2014.921567

Torres, A. (2017). "I am Undocumented and a New Yorker": Affirmative City Citizenship and New York City's IDNYC Program. *Fordham Law Review* 86(1): 335–66.

Tsavdaroglou, C. (2018). The Newcomer's Right to the Common Space: The Case of Athens During the Refugee Crisis. *ACME: An International E-Journal for Critical Geographies* 17(2).

UNICEF (2012). *Children in Immigrant Families in Eight Affluent Countries*. Florence: Innocenti Center.

Unzueta Carrasco, T. A. and Seif, H. (2014). Disrupting the Dream: Undocumented Youth Reframe Citizenship and Deportability through Anti-deportation Activism. *Latino Studies* 12(2): 279–99. https://doi.org/10.1057/lst.2014.21

Valenzuela, A., Gonzalez, A. L., Melendez, E., and Theodore, N. (2006). *On the Corner: Day Labor in the United States*. https://www.issuelab.org/resource/on-the-corner-day-labor-in-the-united-states.html

Vaquera, E., Aranda, E., and Gonzales, R. G. (2014). Patterns of Incorporation of Latinos in Old and New Destinations: From Invisible to Hypervisible. *American Behavioral Scientist* 58(14): 1823–33.

Vargas, J. A. (2011, June 22). My Life as an Undocumented Immigrant. *The New York Times*. https://www.nytimes.com/2011/06/26/magazine/my-life-as-an-undocumented-immigrant.html

Varsanyi, M. (2010). *Taking Local Control: Immigration Policy Activism in US Cities and States*. Redwood City, CA: Stanford University Press.

Vespe, M., Natale, F., and Pappalardo, L. (2017). Data Sets on Irregular Migration and Irregular Migrants in the European Union. *Migration Policy Practice* 7(2): 26–33.

Vogt, W. A. (2013). Crossing Mexico: Structural Violence and the Commodification of Undocumented Central American Migrants. *American Ethnologist* 40(4): 764–80. https://doi.org/10.1111/amet.12053

Vollmer, B. (2008). Undocumented Migration in the UK: Counting the Uncountable: Data and Trends across Europe. *Research Briefs*. https://www.narcis.nl/publication/RecordID/oai:dare.uva.nl:publications%2F97b7fe21-726b-4240-8aac-b8ae52c7ad97

Vollmer, B. (2011). Irregular Migration in the UK: Definitions, Pathways and Scale. *The Migration Observatory*, University of Oxford.

Voss, K. and Bloemraad, I. (2011). *Rallying for Immigrant Rights: The Fight for Inclusion in 21st Century America*. Berkeley, CA: University of California Press.

References

Vradis, A., Papada, E., Painter, J., and Papoutsi, A. (2018) *New Borders: Hotspots and the European Migration Regime*. London: Pluto Press.

Waldinger, R. and Catron, P. (2016). Modes of Incorporation: A Conceptual and Empirical Critique. *Journal of Ethnic and Migration Studies* 42(1): 23–53. https://doi.org/10.1080/1369183X.2015.1113742

Walker, K. E. and Leitner, H. (2011). The Variegated Landscape of Local Immigration Policies in the United States. *Urban Geography* 32(2): 156–78. https://www-tandfonline-com.ezp-prod1.hul.harvard.edu/doi/abs/10.2747/0272-3638.32.2.156

Walters, W. (2008). Acts of Demonstration: Mapping the Territory of (Non-) Citizenship, in E. Isin and G. Neilson (eds), *Acts of Citizenship*. London: Zed Books, pp. 182–206.

Warren, R. and Passel, J. (1987). A Count of the Uncountable: Estimates of Undocumented Aliens Counted in the 1980 United States Census. *Demography* 24(3): 375–93. http://www.jstor.org/stable/2061304

Waters, M. C. and Kasinitz, P. (2013). Immigrants in New York City: Reaping the Benefits of Continuous Immigration. *Daedalus* 142(3): 92–106.

Welch, M., and Schuster, L. (2005). Detention of Asylum Seekers in the US, UK, France, Germany, and Italy: A Critical View of the Globalizing Culture of Control. *Criminal Justice* 5(4): 331–55. https://doi.org/10.1177/1466802505057715

Willen, S. S. (2007). Toward a Critical Phenomenology of "Illegality": State Power, Criminalization, and Abjectivity among Undocumented Migrant Workers in Tel Aviv, Israel. International Migration, 45: 8–38. doi:10.1111/j.1468-2435.2007.00409.x

Winders, J. (2012). Seeing Immigrants: Institutional Visibility and Immigrant Incorporation in New Immigrant Destinations. *Annals of the American Academy of Political and Social Science* 641(1): 58–78.

Wong, J. and Tseng, V. (2008). Political Socialisation in Immigrant Families: Challenging Top-down Parental Socialisation Models. *Journal of Ethnic and Migration Studies* 34(1): 151–68.

Woodbridge, J. (2005). *Sizing the Unauthorised (Illegal) Migrant Population in the United Kingdom in 2001*. London: Home Office.

Wortham, S., Mortimer, K., and Allard, E. (2009). Mexicans as Model Minorities in the New Latino Diaspora. *Anthropology and Education Quarterly* 40(4): 388–404. http://dx.doi.org.ezp-prod1.hul.harvard.edu/10.1111/j.1548-1492.2009.01058.x

Yarris, K. E. (2017). *Care across Generations: Solidarity and Sacrifice in Transnational Families*. Redwood City, CA: Stanford University Press.

Yeo, C. (2018). The Impact of the UK–EU Agreement on Residence Rights for EU Families. Eurochildren Research Brief Series, no. 1: https://eurochildrenblog.files.wordpress.com/2018/03/eurochildren-brief-1-colin-yeo.pdf

Yoshikawa, H. (2011). *Immigrants Raising Citizens: Undocumented Parents*

References

and Their Young Children. New York: Russell Sage Foundation. http://nrs.harvard.edu/urn-3:hul.ebookbatch.PMUSE_batch:muse9781610447072

Yoshikawa, H. and Kalil, A. (2011). The Effects of Parental Undocumented Status on the Developmental Contexts of Young Children in Immigrant Families. *Child Development Perspectives* 5(4): 291–7. https://doi.org/10.1111/j.1750-8606.2011.00204.x

Yoshikawa, H., Suárez-Orozco, C., and Gonzales, R. G. (2017). Unauthorized Status and Youth Development in the United States: Consensus Statement of the Society for Research on Adolescence. *Journal of Research on Adolescence: The Official Journal of the Society for Research on Adolescence* 27(1): 4–19. https://doi.org/10.1111/jora.12272

Yuval-Davis, N. (2004). Borders, Boundaries, and the Politics of Belonging, in S. May, T. Modood, and J. Squires (eds), *Ethnicity, Nationalism, and Minority Rights.* Cambridge: Cambridge University Press, pp. 214–30. doi:10.1017/CBO9780511489235.011

Yuval-Davis, N. (2006). Belonging and the Politics of Belonging. *Patterns of Prejudice* 40(3): 197–214. https://doi.org/10.1080/00313220600769331

Zatz, M. S. and Rodriguez, N. (2015). *Dreams and Nightmares: Immigration Policy, Youth, and Families,* 1st edn. Berkeley, CA: University of California Press.

Zetter, R. (2007). More Labels, Fewer Refugees: Remaking the Refugee Label in an Era of Globalization. *Journal of Refugee Studies* 20(2): 172–92. https://doi.org/10.1093/jrs/fem011

Zhou, M. and Bankston, C. L. (1998). *Growing Up American: How Vietnamese Children Adapt to Life in the United States.* New York: Russell Sage Foundation. http://nrs.harvard.edu/urn-3:hul.ebookbatch.PMUSE_batch:20170204muse34067

Zimmerman, A. M. (2012). Documenting DREAMs: New Media, Undocumented Youth and the Immigrant Rights Movement. Retrieved February 2, 2019, from https://dmlhub.net/publications/documenting-dreams-new-media-undocumented-youth-and-the-immigrant-rights-movement/

Zlolniski, C. (2006). *Janitors, Street Vendors, and Activists: The Lives of Mexican Immigrants in Silicon Valley.* Berkeley, CA: University of California Press.

Zúñiga, V. and Hernández-León, R. (2005). *New Destinations: Mexican Immigration in the United States.* New York: Russell Sage Foundation.

Zúñiga, V. and Hernández-León, R., Shadduck-Hernández, J., and Villarreal, M. O. (2002). The New Paths of Mexican Immigrants in the United States: Challenges for Education and the Role of Mexican Universities. *Education in the New Latino Diaspora: Policy and the Politics of Identity,* 2, 99.

Index

Abdul 10–11
abjectivity 44
Abrego, Leisy 43, 80, 129
advocacy campaigns 117–18
Agamben, Giorgio 44, 149–50, 153
Ahmed 12–13
Alabama
  House Bill 56, 84
Allegations Management System
  (Britain) 63
ambivalence 149–50
Amuedo-Dorantes, C. 160
Anderson, B. 52
anti-immigrant discourse 6, 19, 83–4, 155
Antiterrorism and Effective Death
  Penalty Act (AEDPA) (1996) (US) 87, 88
Arce, Julissa 98, 98–9
Arizona
  Operation Safeguard 89
  "Show me your papers" directive 83–4, 96
Asia
  undocumented migration to US 23, 30
assemblage
  definition 28
  see also illegality assemblage
asylum seekers 25
  current backlash against 17–18
  and detention 93–4

Athens 73–5
  grassroots strategies in helping migrants 75
  increase in migrants since refugee crisis 73–4
  Solidarity Cities initiative 74
  Xenios Zeus operation 63–4
autonomy of migration 149, 153

Baltimore 58
Barcelona 61, 71–3
  changing of registration of residency rules 72
  local ID card initiative 72, 76
  street-vending issue 72–3
Becker, Howard 104
belonging 38, 45–9
  building own spaces of 109
  and citizenship 47–8
  and membership 46
  politics of 48
  and sense of community 46–7
  thin boundary between illegality and 49–54
Bloch, A. 117
Bloemraad, Irene 158, 160
borders/bordering 8, 58–9, 76, 147
  control of 64
  crossing of international 14, 19, 21, 26, 27
  everyday 62–75, 76, 77, 81
  national 38

195

Index

borders/bordering *(cont.)*
  tightening of 14, 16, 147
  US–Mexico *see* US–Mexico border
Bosniak, Linda 37, 50, 150
Boswell, C. 62
Bourdieu, Pierre 116
Britain
  Allegations Management System 63
  Brexit 20, 64–5
  detention/detention centers 92
  education and migrants 111–12
  and family reunification 128
  hostile environment policy 84–6, 96–7
  Johnson's proposed amnesty for undocumented migrants 57
  NHS involvement in immigration checks 81–3
  number of undocumented migrants in 24, 57
  referendum on EU membership 7
  "Windrush scandal" (2018) 85, 97
British Medical Association 82
Burciaga, Edelina 103

Calais Jungle (France) 12, 154
Calavia, Kitty 40–1
California
  legacy of Proposition 187 83–4
  Operation Gatekeeper 89
Cameron, David 63
camps *see* migrant camps; refugee camps
campzenship 154
capitalism 59
Castillo, Bayron Cardona 153
challenging exclusion 144–63
  and ambivalence 149–50
  and citizenship 148–9
  civil disobedience acts 144, 145
  constructing frames of deservingness 156–8
  and deportation 145–6
  development of a legal and political consciousness 148
  DREAMers 9, 118, 145, 157–8
  identity and migrant politicization 155–61
  marching for freedom 150–2
  migrant camps 153–4
  political socialization within the family 159–61
  protests by young people 144–5
  redefining membership 147–54
  rights claims 148–9
  Sans Papiers movement 9, 146, 151
Chase, Elaine 141
Chavez, Leo R. 44, 130
children 159–60
  avoidance behaviors 138
  and DACA program 6, 20, 121, 122–4, 142
  and detention 93, 95–6
  fear of deportability 133
  heavy burdens and tolls on 133, 134, 143
  household responsibilities and roles 134–5, 143
  schooling and the transition to illegality 136–9, 143
  tension with parents 135
  as translators for their parents 112, 134–5, 159–60
  unaccompanied 139–42
  and uneven geographies of entitlement 150
  see also youth, migrant
cities 8, 76
  attraction of migrants 59
  help offered to migrants by local organizations 61
  immigrants as integral to 60
  inequalities 59
  role of local government in integration of immigrants 60–1
citizenship 16, 37, 47–8, 50, 54, 147–9, 156
  active forms of 149
  and belonging 47–8
  as a "collective practice" 146
  cultural 38, 48, 49, 149, 156
  definitions of 47–8
  global 47, 54

Index

hard boundaries and soft interior 37–8
legal 38, 49, 54
multicultural 54, 57
post-national 47, 54, 149
as practice and performance 148
social 38, 47, 54
soft borders of 150
transnational 47, 54
vernacular notions of 48
Clandestino Project (2009) 23
Clara 15
class 102, 119, 149
Clinton, President Bill 87
Colau, Ada 72
Coleman, James 115
community, sense of 46–7
community-institutional context 106–10
Council of Europe 21
Coutin, Susan Bibler 43, 162–3
Criminal Alien Program (CAP) 89–90
criminal justice system 131
criminalization of undocumented migrants 80, 88–9, 90, 94, 126, 155
Cronk, Heather 152
Cruise Europa 1
cultural citizenship 38, 48, 49, 149, 156
Customs and Border Protection (CBP) (US) 88

DACA (Deferred Action for Childhood Arrivals) program 6, 20, 121, 122–4, 142, 145, 156
de Blasio, Bill 65–6
De Genova, Nicholas 39, 41–3, 79–80
Deferred Action for Childhood Arrivals *see* DACA program
Deleuze, Gilles 28
denizens 52
Department of Homeland Security (DHS) (US) 22–3
deportation(s) 14, 41–2, 53, 80, 86–92, 97, 132, 145–6

changes that have widened landscape of 91–2
and Criminal Alien Program (CAP) 89
fear of 79, 117, 133, 143
number of US 92
statistics 86
threat of 41–2, 43, 44
and use of prosecutorial discretion 91
deservingness, constructing frames of 156–8
detention 14, 80, 86, 86–92, 87–8, 92–6, 97
adverse effect on mental well-being 95
of asylum seekers 93–4
of children 93, 95–6
effects on detainees 93
gendered dimensions 95
and immigration industrial complex 94
purposes of 93
sexual abuse in detention centers 95
Development, Relief, and Education for Alien Minors Act (DREAM Act) 144, 145
DREAMer movement 9, 118, 145, 157–8
Dreby, Joanna 129
drivers' licences 20, 34, 51, 69, 70, 91, 105–6, 119, 127
Durbin, Dick 144

education 100, 102, 106
and Greece 110
see also schools/schooling
Egypt 23
enforcement *see* immigration enforcement
Ericson, R. V. 28–9
EU/Europe
asylum applications 24
border controls 64
challenging exclusion 146
detention facilities 93–4
enlargement of 25

Index

EU/Europe (*cont.*)
  and family reunification 128
  lifting of passport control between Schengen-area member states 63
  and March for Freedom (2012) 150–1
  number of deportations 86
  number of undocumented migrants in 23–4
  "refugee crisis" 71, 73, 110, 118, 128, 140
  restrictionist immigration policies 126
  Sans Papiers movement 9, 146, 151
  unaccompanied migrant children 140–1
  undocumented migrant children 133
EUROCITIES network 74
European Commission 23
Eurostat 24
everyday bordering 62–75, 76, 77, 81
exclusion, challenging of *see* challenging exclusion

families 9, 121–43
  avoidance behavior 134
  child–parent tensions 135
  forgoing of services and programs 133–4
  and Hart–Celler Act (1965) 125
  mixed-status 130–6, 142, 143, 160
  negative effects of labor participation on 133
  political socialization within 159–61
  remittances sent back home 129
  reunification 128, 143
  and separation 125–7, 129–30, 132, 142
  strains caused by separation 129–30
  toll and strains of undocumented status 132–6, 143
  transnational 127–30, 142
  see also children
Feliciano, Cynthia 101
Flores, William V. 48

food industry 114
Foucault, M. 44
France
  and immigration detention 94
  Sans Papiers movement 9, 146, 151
friendships 117
Frontex 64
Fusell, Elizabeth 114

gender
  children in the workforce 135
  and detention 95
  and labor migration 128
  and remittances 129
Georgia 105
Germany
  and detention 94
  number of undocumented migrants 23
global citizenship 47, 54
Global Compact for Safe, Orderly and Regular Migration (GCM) 56
globalization 48, 76
  local manifestations of 59–62
Goldring, Luin 52
Gomberg-Muñoz, Ruth 131
Gonzales, Roberto G. 44, 103, 137, 150, 160–1
Gordon, I. 24
Greece 73, 141
  educational system 110
  and refugee crisis 73–4, 110
Guattari, Felix 28
Guiraudon, Virginie 151

Haggerty, K. D. 28–9
Hammar, Thomas 52
Hart–Celler Act (1965) (US) 125
Hatch, Orrin 144
Home Office 24
Hong, Ju 158
Horton, Sarah B. 114
hostile environment policy 80–6
  Britain 84–6, 96–7
  United States 83–4
Hughes, Everett 104
Hughes, V. 111

198

Index

ICE (Immigration and Customs Enforcement) 70, 78, 86, 88, 90
identity
  and migrant politicization 155–61
Illegal Immigration Reform and Immigrant Responsibility Act (IIRAIRA) (US) (1996) 87, 88
illegality assemblage 5, 25–9, 32
illegality, migrant 3–4, 5, 6, 8, 9, 33–56
  and belonging 45–9
  condition of 42–5
  in context 38–45
  as a produced phenomenon 40–2, 54
  rooted in legal classification 17
  shaped by different types of interactions 49
  spaces between legality and 50–4
  thin boundary between belonging and 49–54
Immigrant National Act (INA) (US) 88
immigrant workforce *see* migrant labor/workforce
immigrants
  integration of *see* integration, immigrant
  legal categories and status designations 16–17, 18
  perceived as a threat 60
  selectivity 101–2
Immigration Act (2014) (Britain) 82
immigration controls 3–4, 14, 38, 44
Immigration and Customs Enforcement *see* ICE
immigration enforcement 8–9, 77, 78–97
  creating a hostile environment 80–6, 96–7
  Criminal Alien Program (CAP) 89–90
  deportation *see* deportation(s)
  detention *see* detention
  expansion of after September 11 88
  impact of 79–80
  Prevention through Deterrence practices 89–92
  Secure Communities Program 89, 90, 91
  seldom color-blind 85, 97
  soft measures 80–1, 96
  287(g) agreements 65, 70, 89, 90
immigration industrial complex 94
immigration laws/regulations 17, 41
Immigration and Nationality Act (US) 70, 90
immigration policies 27
immigration status(es) 39, 53, 156
  occupation of gray space between legality and illegality 51–2
integration, immigrant 31, 69, 102, 107
  and local government 60, 61
  and schools 111, 112–13, 120, 134, 161
International Centre for Migration Policy Development (ICMPD) 21
International Organization for Migration (IOM) 3, 21
isolation 117
Italy 40–1, 141, 154
  detention 94
  migration to 30

Johnson, Boris 57
Justice for Janitors 145

Kaminis, Giorgos 74
Kamran 141–2
Kubal, Agnieszka 52, 53

labor market(s) 100, 120
  access to 113–15
labor migrants *see* migrant labor/workforce
legal citizenship 38, 49, 54
legal violence 43
legality
  liminal 52
  semi- 53
Libya 23
liminal legality 52

Index

local government
  role in immigrant integration 60, 61
local policies
  shaping of undocumented migrants' opportunities and experiences 104–6, 119
London 57, 62
London School of Economics 57
Lopez, M. 160
Los Angeles 105, 109

McCain, John 144
McMahon, Simon 15
McNevin, Anne 149–50
March for Freedom (2012) 150–2
Marshall, T. H. 47, 147–8
Martínez, Lulú 145–6
master status concept 103–4
Mateo, Lizbeth 144, 145–6
May, Theresa 57, 62, 84
media coverage 16
Mediterranean
  crossing of by migrants and deaths caused 6, 13–15, 23–4, 126
MEDMIG (Mediterranean Migration Crisis) 15
megacities 59
membership
  and belonging 46
  redefining 147–54
Menjívar, Cecilia 43, 52–3, 62–3, 80, 116
Mexico/Mexicans
  undocumented migration to United States 22, 23, 29, 30, 41, 49, 68
  see also US–Mexico border
migrant camps 153–4
migrant illegality see illegality, migrant
migrant labor/workforce 38, 39, 41, 54, 59–60, 113–15, 119
  deportation threat dynamic 114
  exploitation of undocumented workers 114, 115
  and food industry 114–15
  gendered 128
  negative effect on families 133
  poor working conditions 114, 133
  and unions 115
migrant rights marches (2006) 160
migrant/migration journeys 10–15, 31, 143
  crossing of Mediterranean and deaths caused 6, 13–15, 23–4, 126
  exposure to violence and danger 14
  hardships and obstacles endured 10–14, 15, 31–2
  vulnerability of women 14
migration
  autonomy of 149, 153
  factors for 26
  rise in international 3
  securitization of 86–7
  state perspective 27
mixed-status families 130–6, 142, 143, 160
modes-of-incorporation framework 103
Mohammad 10–13, 15, 31
Morristown (Tennessee, United States)
  ICE raid on 78–9
Mountjoy, Dick 83
multicultural citizenship 47, 54

national contexts
  role of in defining undocumented migrants 29–31
National Fugitive Operations Program (NFOP) 89
Negrón-Gonzales, Genevieve 155, 160
Netherlands
  migrant youth mobilization 157
Nevins, Joseph 17
New York City 61, 62, 65–8, 109
  IDNYC card 66, 76
  immigration history 66–7
  meeting needs of immigrants 67
Ngai, Mae 16
NHS (National Health Services) 82
Nicholls, Walter 157
Nogales (Arizona) 89–90
non-existence, dimensions of 43

Index

Northern Triangle 139–40
notary public 35

Obama, President Barack 57, 78, 91, 145, 158
overstayers 15, 20, 22, 23, 24, 26, 29, 102, 131

Parisi, Rosa 154
Passel, Jeffrey 21–2
Patras (Greece)
  attempt by migrants to board trucks at 1–3
Plyler v. Doe (1982) 111
political socialization
  within the family 159–61
politicization, migrant 149, 153, 155–61
Portes, Alejandro 102
post-national citizenship 47, 54, 149
poverty 19, 69, 108, 116–17, 129, 132, 154
pre-migration factors 101–2, 104, 119
precarization of rights 17–18
Probyn, Elspeth 46
prosecutorial discretion 91

racial segregation 109
racism 109
Rawlings-Blake, Stephanie 58
reception 67, 103, 107, 108, 119
refuge, seeking safe 152–4
refugee camps 73, 75, 93, 153
refugee crisis (2015) 71, 73, 110, 118, 128, 140
refugees 26
  current backlash against 17–18
  and precarization of rights 17–18
remittances 129
rights claims 148–9
Roma settlements (Italy) 154
Romney, Mitt 81
Rosaldo, Renato 48
Ruhs, M. 52
rural areas

as new immigrant destinations 108–9
Rygiel, K. 154

Saavedra, Marcos 145–6
Salter, Mark 29
Salvadorans 117
Samaniego, Eduardo 161–3
sanctuary cities 61, 67, 71
Sanctuary movement 118, 145
Sans Papiers movement 9, 146, 151
Sarason, Seymour 46
Sassen, Saskia 28, 59
schools/schooling 110–13, 120, 136–7
  Britain 111–12
  Greece 110
  important role in integration process 111, 112–13, 120, 161
  role in students' political and civic development 161
Schuster, Liza 95
Secure Communities 89, 90, 91
securitization of migration 86–7
semi-legality 53
separation
  and families 125–7, 129–30, 132, 142
September 11 (2001) 88
Sergio 33–7, 55
Sessions, General Jeff 6, 121, 122
Showing Up for Racial Justice 152
Sigona, Nando 15, 64–5, 111, 141, 150, 154
Silverman, Stephanie 95
social capital 115–16
social citizenship 47, 54
social mobility 98–120
  and connections 115–18
  and education/schools 110–13, 120
  impact of local policies on 104–6, 119
  and labor markets 113–15, 120
  pre-migration factors 101–2, 104, 119
  and social capital 115–16
social movements, immigrant 118, 144

201

Index

social networks 116–18
social psychology 45, 46
sociological theory 45–6
soft enforcement 80–1, 96
South Carolina 68–71
  challenges faced by undocumented migrants in 69–70
  denial of access to higher education for undocumented migrants 70
  formal and informal forms of support for undocumented migrants 71
  and 287(g) agreements 70
Southern California drywall strike 145
Spain 40–1
  migrants and local politics of belonging in Barcelona 71–3
  migration to 30
Spencer, Sarah 150
Stuesse, Angela
Scratching Out a Living 114
suburbs
  as new immigrant destinations 108–9
support networks 109, 141
Sweden 141
  REVA project 64
Swerts, Thomas 151
Syrian civil war 23

Temporary Protected Status *see* TPS
Terriquez, Veronica 159
Texas
  Operation Hold the Line 89
Torres, Adelita 122–4
TPS (Temporary Protected Status) 6, 20, 30–1, 52, 156
transnational citizenship 47, 54
transnational families 127–30, 142
Trump, President Donald 91
  attempt to end TPS 30–1
  curtailing of immigration 6, 18
Tseng, V. 159
287(g) agreements 65, 70, 89, 90

UN High Commissioner for Refugees (UNHCR) 14, 23
unaccompanied migrant youth 139–42
undocumented migrants/migration
  access to rights and resources 18–19
  becoming process 20, 32, 48
  challenges faced by 119
  in a changing world 15–19
  countries come from 14
  definition 19–20
  estimating number of 21–5
  evasive nature of 21, 32, 38
  exclusions experienced 104
  lived experience of 33–56
  motives 15, 19, 26
  non-existence dimensions 43
  persistence of 16
  role of national contexts in defining 29–31
  structuring of opportunities by local contexts 102–18
  subjected to excessive forms of policing and denial of rights 42, 55
undocumented status 98–120
unions 115
United States
  anti-immigrant legislation 83–4
  Criminal Alien Program (CAP) 89–90
  curtailing of immigration by Trump 6, 18
  DACA program 6, 20, 121, 122–4, 142, 145, 156
  deportation 86, 87–8, 91, 92, 145–6
  detention/detention centres 92–3, 94, 95
  family reunification 130
  immigrant labor in food industry 114–15
  immigration enforcement measures 87–92
  immigration laws 41
  local law enforcement units and immigration enforcement 60–1

Index

local policies towards immigrants 105
Mexican migrants 22, 23, 29, 30, 41, 49, 68
  and migrant labor 41
  migrant-rights marches (2006) 160
  mixed-status families 130–1
  number of undocumented migrants in 21–3
  Obama's immigration reform attempt 57–8
  relationship between state and federal governments 76–7
  schooling for migrant children 136
  Secure Communities Program 90
  shift of Latin American migration from cities to rural destinations 68–9
  Temporary Protected Status (TPS) 6, 20, 30–1, 52, 156
  287(g) agreements 65, 70, 89, 90
  undocumented migrants in 29–30, 83, 125
  use of prosecutorial discretion 91
  youth mobilization 157
urbanization 59, 59–60
US–Mexico border 65, 83, 87, 89–90, 93, 139, 140, 145–6, 152
  arrival of immigrant caravan (2018) 152–3

Vargas, Jose Antonio 98, 99–100
Vietnamese 116
Wacquant, Loic 116
Walters, W. 153
Welch, Michael 95
Willen, Sarah 43–4
Wilson, Pete 83
"Windrush scandal" (2018) 85, 97
women
  and detention 95
  migration of 128, 129
  vulnerability of during migration journeys 14
Wong, J. 159
Woodbridge, Jo 24
worker centers 115

youth, migrant 155–6
  challenges faced by 136–7
  constructing frames of deservingness 157–8
  political socialization within the family 159–61
  protests by and political mobilization 144, 155–6, 157–8
  and transition to illegality 136–9
  unaccompanied migrant 139–42
Yuval-Davis, Nira 48

Zetter, Roger 17
Zhou, Min and Bankston, Carol
  Growing Up American 116